REGRESSION AND THE MOORE–PENROSE PSEUDOINVERSE

This is Volume 94 in
MATHEMATICS IN SCIENCE AND ENGINEERING
A series of monographs and textbooks
Edited by RICHARD BELLMAN, *University of Southern California*

The complete listing of books in this series is available from the Publisher upon request.

REGRESSION AND THE MOORE-PENROSE PSEUDOINVERSE

Arthur Albert

DEPARTMENT OF MATHEMATICS
BOSTON UNIVERSITY
BOSTON, MASSACHUSETTS

1972

ACADEMIC PRESS New York and London

ACADEMIC PRESS, INC.
111 Fifth Avenue, New York, New York 10003

United Kingdom Edition published by
ACADEMIC PRESS, INC. (LONDON) LTD.
24/28 Oval Road, London NW1

LIBRARY OF CONGRESS CATALOG CARD NUMBER: 72-77337

AMS (MOS) 1970 Subject Classifications: 62J05, 15A09

PRINTED IN THE UNITED STATES OF AMERICA

To my Father and Mother
on their anniversary

CONTENTS

PREFACE xi
ACKNOWLEDGMENTS xiii

Part I THE GENERAL THEORY AND COMPUTATIONAL METHODS

Chapter I. **Introduction** 3

Chapter II. **General Background Material**

2.1–2.9 Linear Manifolds in Finite-Dimensional Euclidean Spaces 6
2.10–2.13 Range Spaces and Null Spaces for Matrices 11
2.14 The Diagonalization Theorem for Symmetric Matrices 12

Chapter III. **Geometric and Analytic Properties of the Moore–Penrose Pseudoinverse**

3.1 Existence of Minimum Norm Solution to Least Squares Minimization 15
3.2 Uniqueness 17
3.3–3.4 $H^{+} = \lim_{\delta\to 0}(H^{\mathrm{T}}H+\delta^{2}I)^{-1}H^{\mathrm{T}}$ 19
3.5 HH^{+} and $H^{+}H$ Are Projections 20
3.6 Specialization to Symmetric H 23
3.7 Exercises 24
3.8 Properties of H^{+}, $H^{+}H$, HH^{+}, etc. 26
3.9 Penrose's Characterization of H^{+} 28
3.10–3.11 Exercises 29
3.12–3.14 Characterization of All Solutions to Least Squares Minimization with Applications to Theory of Linear Equations, Constrained Least Squares, Linear Programming, Projections, and Markov Chains 30

3.15–3.18 Singular Value Decomposition Theorems 38
3.19 Iterative Method for Finding Dominant Eigenvalue and Eigenvector for AA^{T} and $A^{\mathrm{T}}A$ 42

Chapter IV. **Pseudoinverses of Partioned Matrices and Sums and Products of Matrices**

4.1–4.3 Partioned Matrices, I 43
4.4–4.5 Application to Stepwise Regression, I 44
4.6 Relationship between $(\sum_{j=1}^{m} c_j c_j^{\mathrm{T}})^+$ and $(\sum_{j=1}^{m+1} c_j c_j^{\mathrm{T}})^+$ 47
4.7 Partioned Matrices, II 49
4.8 $(UU^{\mathrm{T}}+VV^{\mathrm{T}})^+$ 49
4.9 Perturbation Theorems 50
4.10 The Concept of Rank 52
4.11–4.12 Pseudoinverse of Products 53
4.13–4.15 Exercises 56

Chapter V. **Computational Methods**

5.1 Gramm–Schmidt Method of Rust, Burrus, and Schneeburger 57
5.2 Gauss–Jordan Elimination Method of Ben-Israel and Wersan and Noble 65
5.3 Gradient Projection Method of Pyle 69
5.4 Cayley–Hamilton Method of Decell, Ben-Israel, and Charnes 74

Part II **STATISTICAL APPLICATIONS**

Chapter VI. **The General Linear Hypothesis**

6.1 Best Linear Unbiased Estimation; The Gauss–Markov Theorem 86
6.2 Distribution for Quadratic Forms in Normal Random Variables 93
6.3 Estimable Vector Parametric Functions and Confidence Ellipsoids in the Case of Normal Residuals 97
6.4 Tests of the General Linear Hypothesis 100
6.5 The Relationship between Confidence Ellipsoids for Gx and Tests of the General Linear Hypothesis 102
6.6 Orthogonal Designs 104

Chapter VII. **Constrained Least Squares, Penalty Functions, and BLUE's**

7.1 Penalty Functions 119
7.2 Constrained Least Squares as Limiting Cases of BLUE's 122

Chapter VIII. **Recursive Computation of Least Squares Estimators**

8.1 Unconstrained Least Squares 125
8.2 Recursive Computation of Residuals 129
8.3 Weighted Least Squares 131
8.4 Recursive Constrained Least Squares, I 133
8.5 Recursive Constrained Least Squares, II 135
8.6 Additional Regressors, II (Stepwise Regression) 146
8.7 Relationship between Analysis of Variance and Analysis of Covariance 148
8.8 Missing Observations 151

Chapter IX. **Nonnegative Definite Matrices, Conditional Expectation, and Kalman Filtering**

9.1 Nonnegative Definiteness 157
9.2 Conditional Expectations for Normal Random Variables 161
9.3 Kalman Filtering 169
9.4 The Relationship between Least Squares Estimates and Conditional Expectations 171

References 173

INDEX 177

PREFACE

For the past ten years, my professional interests have focused on various aspects of regression. It has been my experience that the pseudoinverse is a great unifying concept. It has helped me to understand, remember, and explain many classical results in statistical theory as well as to discover (and rediscover) some new ones.

This book was conceived as a hybrid monograph–textbook. As a text, it would be suitable for the equivalent of a two-quarter course. In teaching such a course, one could fill in the remainder of the year with additional material on (for example) multiple regression, nonlinear regression, large sample theory, and optimal experimental design for regression. For this purpose I have attempted to make the development didactic. On the other hand, most of the material comes from reasonably current periodical literature and a fair amount of the material is my own work (some already published, some not). Virtually all of the material deals with regression either directly (Chapters VI–IX) or as background (Chapters I–V). By restricting the domain of discourse we are able to pursue a leisurely pace and, I hope, to preserve a sense of unity throughout.

At the time the manuscript was completed there were, to my knowledge, no textbook treatments of the pseudoinverse. Since that time, two excellent complementary monographs have appeared containing treatments of the Moore–Penrose pseudoinverse in a more general setting. The first (Boullion and Odell [2]) appeared in early 1971 and concerns itself mainly with algebraic and structural properties of these pseudoinverses. The second (Rao and Mitra [1]) appeared later in 1971 and is extremely comprehensive in its coverage of the then extant pseudoinverse literature. Both volumes contain large bibliographies.

ACKNOWLEDGMENTS

I wish to express my thanks to the Office of Naval Research, Army Research Office, and Air Force Office of Scientific Research for their support during various stages of this book. In particular, the final draft was written while I was a visitor at Stanford during the summer of 1970, at which time I was supported under contract N00014–67–A–0112–0053; NR–042–267.

The index for this book was sorted and alphabetized on our time-share computer terminal. I would like to thank Don Feinberg, who wrote the program, and Nancy Burkey, who fed the necessary information to the computer.

Part I

THE GENERAL THEORY AND COMPUTATIONAL METHODS

Chapter I

INTRODUCTION

In 1920, Moore [1] introduced the notion of a generalized inverse for matrices. The idea apparently lay dormant for about 30 years, whereupon a renaissance of interest occurred. As may be appreciated after a brief look through the bibliography, a large and varied literature has appeared since the early 1950s.

At the present time, the theory is elegant, the applications are diverse (e.g., least squares, linear equations, projections, statistical regression analysis, filtering, and linear programming) and most important, a deeper understanding of these topics is achieved when they are studied in the generalized inverse context. The results that we present here are, for the most part, scattered throughout the periodical literature or to be found in out-of-print technical reports. This is the major impetus behind the writing of this monograph.

The level of the material presupposes a familiarity with the notion of "limit" and some of the fundamental properties of finite-dimensional Euclidean spaces. That much will suffice for the first half of the book. The second half is devoted to statistical applications and for these it would be helpful to have had a first course in probability (and/or statistics).

Many exercises are included. (Solution outlines are provided in a separate booklet.) Some of the exercises are supplementary to the material in the monograph, whereas others are lemmas to subsequent theorems. The reader is urged to do the exercises. They are the key to a thorough understanding of the material.

Sections and equations are numbered according to a decimal system. For example, Eq. (6.3.1) comes after Section (6.2) and before Definition (6.3.1.9). The latter comes before Eq. (6.4). Every effort has been made to maintain the following typographical conventions:

Sets—uppercase script
Matrices—uppercase Latin
Vectors—lowercase Latin
Scalars—lowercase Greek
Vector random variables—lowercase boldface Latin
Real random variables—lowercase boldface Greek.

Bibliographic references are enclosed in square brackets; equation and section numbers are written in parentheses at the left. Sometimes a section or an equation number appears in the middle of a line or at the right side. In this case, the assertion preceding this equation number is a direct consequence of that (previously established) result.

Chapter II

GENERAL BACKGROUND MATERIAL

In this chapter, we will review the key results and definitions from the theory of real linear spaces which are relevant to what follows. This survey is informal and presumes previous exposure of the reader to these notions. (For a more formal review, see Appendix A of Karlin [1]. For a rigorous development, see Halmos [1] or the first half of Bellman [1].)

We begin by reminding the reader that linear transformations from one Euclidean space to another can be represented by matrices, once the coordinate systems for the two spaces have been decided upon. Furthermore, vectors can be represented as long, skinny matrices (having one column). If A is any matrix, we denote the transpose of A by A^{T}. In our discussions, all matrices, vectors (and scalars) have real entries. Matrix transposition has the following important properties:

$$(AB)^{\mathrm{T}} = B^{\mathrm{T}}A^{\mathrm{T}}, \qquad (A^{\mathrm{T}})^{\mathrm{T}} = A, \qquad (A+B)^{\mathrm{T}} = A^{\mathrm{T}}+B^{\mathrm{T}}.$$

If x and y are two vectors having the same number of components, the *inner product* (or scalar product) of x with y is the scalar $x^{\mathrm{T}}y$ (which is the same as $y^{\mathrm{T}}x$). The *norm* of x is $\|x\| = (x^{\mathrm{T}}x)^{1/2}$. Two vectors are said to be *orthogonal* if their inner product vanishes. (The statement, x is orthogonal to y is abbreviated $x \perp y$.)

A *linear manifold* $\mathscr{L}$ is a nonempty subset of a Euclidean space, which is closed under addition and scalar multiplication (if x and y are elements of $\mathscr{L}$ then for any scalars α and β, $\alpha x+\beta y$ are members of $\mathscr{L}$).

A vector x is *orthogonal to the linear manifold* $\mathscr{L}$ if x is orthogonal to every vector in $\mathscr{L}$ (abbreviated $x \perp \mathscr{L}$).

The symbol $\in$ will be reserved for set membership ($x \in \mathscr{L}$ means that x is a member of the set $\mathscr{L}$). The symbol $\subseteq$ denotes set inclusion and $\subset$ denotes proper inclusion.

The following theorem is of fundamental importance in all that follows. We state it without proof:

(2.1) Theorem: Let x be a vector in a finite-dimensional Euclidean space and let $\mathscr{L}$ be a linear manifold in that space. Then there is a unique vector $\hat{x} \in \mathscr{L}$ having the property that $x-\hat{x} \perp \mathscr{L}$.

(2.1.1) *Comment:* An equivalent statement of the theorem is that there is a unique decomposition of x:

$$x = \hat{x} + \tilde{x}$$

where

$$\hat{x} \in \mathscr{L} \qquad \text{and} \qquad \tilde{x} \perp \mathscr{L}.$$

The vector $\hat{x}$ is called the *projection of x on $\mathscr{L}$*. It is the vector in $\mathscr{L}$ that is "nearest" to x, as we shall now demonstrate.

(2.2) Theorem: Let x be a vector and $\mathscr{L}$ a linear manifold. If $x=\hat{x}+\tilde{x}$ where $\hat{x} \in \mathscr{L}$ and $\tilde{x} \perp \mathscr{L}$, then

$$\|x-y\| > \|x-\hat{x}\|$$

if

$$y \in \mathscr{L} \qquad \text{and} \qquad y \neq \hat{x}.$$

Proof: If $y \in \mathscr{L}$ then

$$\begin{aligned}\|x-y\|^2 = \|\hat{x}+\tilde{x}-y\|^2 &= \|(\hat{x}-y)+\tilde{x}\|^2 \\ &= \|\hat{x}-y\|^2 + \|\tilde{x}\|^2\end{aligned}$$

since $\tilde{x} \perp \mathscr{L}$ and $\hat{x}-y \in \mathscr{L}$. Therefore

$$\|x-y\|^2 \geqslant \|\tilde{x}\|^2$$

with strict inequality holding unless $\|\hat{x}-y\|^2 = 0$. ■

Theorems (2.1) and (2.2) are "existence theorems." As a consequence of the next theorem, we can show how to reduce the computation of $\hat{x}$ to the solution of simultaneous linear equations. First though, we remind the reader that a linear manifold $\mathscr{L}$ *is spanned by* $(y_1, y_2, \ldots, y_n)$ if every vector in $\mathscr{L}$ can be expressed as a linear combination of the y_j's.

(2.3) Theorem: (a) If x is a vector and $\mathscr{L}$ is a linear manifold, then $\hat{x}$, the projection of x on $\mathscr{L}$, is the unique vector in $\mathscr{L}$ satisfying the equations

(2.3.1) $$\hat{x}^{\mathrm{T}} y = x^{\mathrm{T}} y \qquad \text{for all} \quad y \in \mathscr{L}.$$

(b) If $\mathscr{L}$ is spanned by $y_1, y_2, \ldots, y_n$, $\hat{x}$ is the unique vector in $\mathscr{L}$ satisfying

(2.3.2) $$\hat{x}^{\mathrm{T}} y_j = x^{\mathrm{T}} y_j \qquad j = 1, 2, \ldots, n.$$

Proof: Part (a) follows directly from (2.1). That $\hat{x}$ satisfies (2.3.2) is a consequence of (a). If x^* is some other vector in $\mathscr{L}$ satisfying

$$x^{*\mathrm{T}} y_j = x^{\mathrm{T}} y_j \qquad j = 1, 2, \ldots, n,$$

then

(2.3.3) $$(x^* - \hat{x})^{\mathrm{T}} y_j = 0 \qquad j = 1, 2, \ldots, n.$$

Since the y_j's span $\mathscr{L}$, any vector in $\mathscr{L}$ is a linear combination of the y_j's, so that $x^* - \hat{x}$ is orthogonal to *every* vector in $\mathscr{L}$. Since $x^* - \hat{x} \in \mathscr{L}$, it follows that $(x^* - \hat{x})^{\mathrm{T}}(x^* - \hat{x}) = \|x^* - \hat{x}\|^2 = 0$. Therefore, x^* must coincide with $\hat{x}$ if it lies in $\mathscr{L}$ and satisfies (2.3.2). ■

(2.4) Exercise: If $\mathscr{L}$ is spanned by $y_1, y_2, \ldots, y_n$, then $\hat{x}$ is the unique vector of the form

$$\hat{x} = \sum_{j=1}^{n} \alpha_j y_j$$

where the α_j's are any scalars satisfying the simultaneous set of linear equations

(2.4.1) $$\sum_{j=1}^{n} \alpha_j (y_i^{\mathrm{T}} y_j) = y_i^{\mathrm{T}} x \qquad i = 1, 2, \ldots, n.$$

(2.5) Exercises

(2.5.1) If $\mathscr{L}$ is spanned by $y_1, y_2, \ldots, y_n$ then $x \perp \mathscr{L}$ if x is orthogonal to each of the y_j's.

(2.5.2) If x and y are vectors in the same Euclidean space, and $\mathscr{L}$ is a linear manifold in that space, then the projection of $\alpha x + \beta y$ on $\mathscr{L}$ is $\alpha\hat{x} + \beta\hat{y}$ where $\hat{x}$ and $\hat{y}$ are the projections of x and y on $\mathscr{L}$.

If $y_1, y_2, \ldots, y_n$ is a set of vectors in the same Euclidean space, we denote *the linear manifold spanned by the* y_j*'s* by $\mathscr{L}(y_1, y_2, \ldots, y_n)$. This manifold is the smallest manifold containing all the y_j's and it consists, exactly, of all those vectors which are expressible as a linear combination of the y_j's.

(2.5.3) The projection of x on $\mathscr{L}(y)$ is $(x^{\mathrm{T}}y)y/\|y\|^2$ if $y \neq 0$.

If $x, y_1, y_2, \ldots, y_n$ is an arbitrary set of vectors in the same Euclidean space, a particularly simple relationship exists between $\hat{x}_n$, the projection of x on $\mathscr{L}(y_1, \ldots, y_n)$, and $\hat{x}_{n-1}$, the projection of x on $\mathscr{L}(y_1, \ldots, y_{n-1})$, provided that y_n is orthogonal to the previous y_j's:

(2.6) Theorem: If $x, y_1, \ldots, y_n$ are vectors in the same Euclidean space and if $y_n \perp \mathscr{L}(y_1, \ldots, y_{n-1})$ then

(2.6.1)
$$\hat{x}_n = \hat{x}_{n-1} + \begin{cases} 0 & \text{if } y_n = 0 \\ (x^{\mathrm{T}}y_n)y_n/\|y_n\|^2 & \text{otherwise.} \end{cases}$$

Proof: Since $\hat{x}_{n-1} \in \mathscr{L}(y_1, \ldots, y_{n-1})$, $\hat{x}_n$ is clearly a member of $\mathscr{L}(y_1, \ldots, y_n)$. It is readily verified that the right side of (2.6.1) satisfies (2.3.2), provided y_n is orthogonal to $\mathscr{L}(y_1, \ldots, y_{n-1})$ (and in particular, orthogonal to $\hat{x}_{n-1}$). The conclusion follows from (2.3b). ■

As an immediate consequence we can derive the so-called Fourier expansion theorem:

(2.7) Theorem: If $u_1, u_2, \ldots, u_n$ are mutually orthogonal vectors of unit length and x is an arbitrary vector in the same Euclidean space, then $\hat{x}$, the projection of x on $\mathscr{L}(u_1, \ldots, u_n)$ is given by

(2.7.1)
$$\hat{x} = \left(\sum_{j=1}^{n} u_j u_j^{\mathrm{T}}\right)x,$$

(2.7.2)
$$\hat{x} = \sum_{j=1}^{n} (u_j^{\mathrm{T}}x)\,u_j.$$

Comment: If the u_j's are k-dimensional vectors, the expression $\sum_{j=1}^{n} u_j u_j^{\mathrm{T}}$ is a $k \times k$ matrix, so that (2.7.1) is a representation of the matrix which projects x onto $\mathscr{L}(u_1, \ldots, u_n)$. Equation (2.7.2), on the other hand, is an explicit representation of $\hat{x}$ as a linear combination of the u_j's.

If $\mathscr{L}_1$ and $\mathscr{L}_2$ are linear manifolds and $\mathscr{L}_1 \subseteq \mathscr{L}_2$, define $\mathscr{L}_2 - \mathscr{L}_1$ as the set of vectors in $\mathscr{L}_2$ which are orthogonal to $\mathscr{L}_1$.

(2.7.3) **Exercise:** If $\mathscr{L}_1 \subseteq \mathscr{L}_2$, then $\mathscr{L}_2 - \mathscr{L}_1$ is a linear manifold.

(2.7.4) **Exercise:** Let x be a vector and suppose $\mathscr{L}_1 \subseteq \mathscr{L}_2$. Define $\hat{x}_2$ to be the projection of x on $\mathscr{L}_2$, $\hat{x}_{21}$ to be the projection of $\hat{x}_2$ on $\mathscr{L}_1$, $\hat{x}_1$ to be the projection of x on $\mathscr{L}_1$ and $\tilde{x}_{21}$ to be the projection of x on $\mathscr{L}_2 - \mathscr{L}_1$.

Then

(a) $\hat{x}_{21} = \hat{x}_1$. (The projection of x on $\mathscr{L}_1$ is obtainable by projecting x on $\mathscr{L}_2$ and then projecting that vector on $\mathscr{L}_1$).

(b) $\hat{x}_2 = \hat{x}_1 + \tilde{x}_{21}$.

(c) $\|x - \hat{x}_1\| \geqslant \|x - \hat{x}_2\|$ with strict inequality holding if $\mathscr{L}_1$ is a proper subset of $\mathscr{L}_2$ unless $x \in \mathscr{L}_1$.

(2.8) The Gram–Schmidt Orthogonalization Procedure

This procedure takes an arbitrary collection of vectors $h_1, h_2, \ldots, h_n$ and generates a set of mutually orthogonal vectors $u_1, u_2, \ldots, u_n$, having the properties that

(2.8.1) $$\mathscr{L}(u_1, u_2, \ldots, u_j) = \mathscr{L}(h_1, h_2, \ldots, h_j) \qquad \text{for} \quad j = 1, \ldots, n$$

and

(2.8.2) $$\|u_j\| = 1 \qquad \text{if} \quad u_j \neq 0 \quad j = 1, 2, \ldots, n.$$

The Procedure

(2.8.3) $$u_1 = \begin{cases} h_1/\|h_1\| & \text{if} \quad h_1 \neq 0 \\ 0 & \text{if} \quad h_1 = 0. \end{cases}$$

For $j = 1, 2, \ldots, n-1$, define

(2.8.4) $$\hat{h}_{j+1} = \sum_{k=1}^{j} (h_{j+1}^T u_k) u_k$$

and

(2.8.5) $$u_{j+1} = \begin{cases} (h_{j+1} - \hat{h}_{j+1})/\|h_{j+1} - \hat{h}_{j+1}\| & \text{if} \quad h_{j+1} - \hat{h}_{j+1} \neq 0 \\ 0 & \text{otherwise}. \end{cases}$$

The properties (2.8.1) and (2.8.2) are established by induction: The induction hypothesis is that

$$\mathscr{L}(u_1, u_2, \ldots, u_j) = \mathscr{L}(h_1, \ldots, h_j)$$

and

$$u_{j+1} \perp \mathscr{L}(u_1, \ldots, u_j).$$

By definition of u_1, $\mathscr{L}(u_1) = \mathscr{L}(h_1)$ and by (2.5.3), $\hat{h}_2$ is the projection of h_2 on $\mathscr{L}(u_1)$. By (2.1), $h_2 - \hat{h}_2 \perp \mathscr{L}(u_1)$ so that u_2 is orthogonal to $\mathscr{L}(u_1)$. This establishes the induction hypothesis for $j = 1$. If it is assumed that the hypothesis is true for all values of j up through k, then since u_{k+1} is a linear combination of h_{k+1} and $\hat{h}_{k+1}$ [which lies in $\mathscr{L}(u_1,\ldots,u_k) = \mathscr{L}(h_1,\ldots,h_k)$], we see that any vector which is a linear combination of $u_1,\ldots,u_{k+1}$ is also a linear combination of $h_1,\ldots,h_{k+1}$. This means that $\mathscr{L}(u_1,\ldots,u_{k+1}) \subseteq \mathscr{L}(h_1,\ldots,h_{k+1})$.

On the other hand,

$$h_{k+1} = \|h_{k+1} - \hat{h}_{k+1}\| u_{k+1} + \hat{h}_{k+1}$$

the right side being a linear combination of the nonzero members of $\{u_1, u_2, \ldots, u_{k+1}\}$. Since $\mathscr{L}(h_1,\ldots,h_k) = \mathscr{L}(u_1,\ldots,u_k)$ under the induction hypothesis, any vector which is expressible as a linear combination of $h_1,\ldots h_{k+1}$ is also expressible as a linear combination of $u_1,\ldots,_{k+1}$. Therefore, $\mathscr{L}(h_1,\ldots,h_{k+1}) \subseteq \mathscr{L}(u_1,\ldots,u_{k+1})$. This establishes the first half of the induction hypothesis for $j = k+1$.

The second half follows since $\hat{h}_{k+2}$ is the projection of h_{k+2} on $\mathscr{L}(u_1,\ldots,u_{k+1})$, (2.7.2). By (2.1), $h_{k+2} - \hat{h}_{k+2} \perp \mathscr{L}(u_1,\ldots,u_{k+1})$ and therefore, so is u_{k+2}. ■

(2.8.6) **Exercise:** $u_j = 0$ if and only if h_j is a linear combination of $(h_1,\ldots,h_{j-1})$.

In what follows, two special linear manifolds will be of particular interest: If H is any matrix, the *null space of* H, denoted by $\mathscr{N}(H)$, is the set of vectors which H maps into zero:

$$\mathscr{N}(H) = \{x\colon Hx = 0\}.$$

[The null space of H, $\mathscr{N}(H)$, always has at least one element, namely the null vector 0.]

The *range of* H, denoted by $\mathscr{R}(H)$, is the set of vectors which are afterimages of vectors in the Euclidean space which serves as the domain of H:

$$\mathscr{R}(H) = \{z\colon z = Hx \quad \text{for some} \quad x.\}$$

It is easy to see that $\mathscr{N}(H)$ and $\mathscr{R}(H)$ are linear manifolds.

(2.9) Exercises

(2.9.1) Let the column vectors of H be denoted by $h_1, h_2, \ldots, h_n$. Show that $\mathscr{R}(H) = \mathscr{L}(h_1, h_2, \ldots, h_n)$.

(2.9.2) Show that H^{T} is the *adjoint* of H. That is to say, if H is an $n \times m$ matrix, then for any m-dimensional vector, x and any n-dimensional vector y, the inner product of x with Hy is the same as the inner product of y with $H^{\mathrm{T}}x$.

If $\mathscr{L}$ is a linear manifold in a Euclidean space $\mathscr{E}$, the *orthogonal complement* of $\mathscr{L}$ (denoted by $\mathscr{L}^{\perp}$) is defined to be the set of vectors in $\mathscr{E}$ which are (each) orthogonal to $\mathscr{L}$.

It is easy to see that $\mathscr{L}^{\perp}$ is itself a linear manifold.

(2.9.3) $(\mathscr{L}^{\perp})^{\perp} = \mathscr{L}$.

(2.9.4) If x is a vector in $\mathscr{E}$ and $x^{\mathrm{T}}y = 0$ for all $y \in \mathscr{E}$, then $x = 0$.

The null space of a matrix is related to the range space of its transpose. In fact, the next theorem shows that the null space of H consists of those vectors which are orthogonal to the column vectors of H^{T} (i.e., the rows of H) which is just another way of saying

(2.10) Theorem: For any matrix H, $\mathscr{N}(H) = \mathscr{R}^{\perp}(H^{\mathrm{T}})$.

Proof: $x \in \mathscr{N}(H)$ if and only if $Hx = 0$. Therefore, $x \in \mathscr{N}(H)$ if and only if $y^{\mathrm{T}}Hx = 0$ for all y (having the correct number of components, of course). [Use (2.9.4).] Since $y^{\mathrm{T}}Hx = (H^{\mathrm{T}}y)^{\mathrm{T}}x$, we see that $Hx = 0$ if and only if x is orthogonal to all vectors of the form $H^{\mathrm{T}}y$. These vectors, collectively, make up $\mathscr{R}(H^{\mathrm{T}})$, thus proving the assertion. ■

By applying the projection theorem (2.1), we deduce as an immediate consequence of (2.10), that every vector z (having the correct number of components) has a unique decomposition as the sum of two terms, one lying in $\mathscr{R}(H)$ and one lying in $\mathscr{N}(H^{\mathrm{T}})$:

(2.11) Theorem: If H is an $n \times m$ matrix and z is an n-dimensional vector, we can uniquely decompose z:

$$z = \hat{z} + \tilde{z}$$

where $\hat{z} \in \mathscr{R}(H)$ and $\tilde{z} \in \mathscr{N}(H^{\mathrm{T}})$.

(2.11.1) **Exercise:** In (2.11), $\hat{z}$ is the projection of z on $\mathscr{R}(H)$ and $\tilde{z}$ is the projection of z on $\mathscr{N}(H^{\mathrm{T}})$. Consequently $H^{\mathrm{T}}z = H^{\mathrm{T}}\hat{z}$.

A matrix is said to be *symmetric* if it is equal to its transpose. Obviously, symmetric matrices are square.

(2.11.2) **Exercise:** Matrices of the form $H^{\mathrm{T}}H$ and HH^{T} are always symmetric.

By virtue of (2.10),

(2.11.3) $$\mathscr{N}(A) = \mathscr{R}^{\perp}(A) \quad \text{and} \quad \mathscr{R}(A) = \mathscr{N}^{\perp}(A)$$

if A is symmetric.

Moreover, if H is any matrix, then

(2.12) Theorem: $\mathscr{R}(H) = \mathscr{R}(HH^{\mathrm{T}})$, $\mathscr{R}(H^{\mathrm{T}}) = \mathscr{R}(H^{\mathrm{T}}H)$, $\mathscr{N}(H) = \mathscr{N}(H^{\mathrm{T}}H)$, and $\mathscr{N}(H^{\mathrm{T}}) = \mathscr{N}(HH^{\mathrm{T}})$.

Proof: It suffices to prove that $\mathscr{N}(H^{\mathrm{T}}) = \mathscr{N}(HH^{\mathrm{T}})$ and $\mathscr{N}(H) = \mathscr{N}(H^{\mathrm{T}}H)$. Then apply (2.10) and (2.9). To prove that $\mathscr{N}(H^{\mathrm{T}}) = \mathscr{N}(HH^{\mathrm{T}})$, we note that $HH^{\mathrm{T}}x = 0$ if $H^{\mathrm{T}}x = 0$. On the other hand, if $HH^{\mathrm{T}}x = 0$, then $x^{\mathrm{T}}HH^{\mathrm{T}}x = 0$ so that $\|H^{\mathrm{T}}x\|^2 = 0$ which implies that $H^{\mathrm{T}}x = 0$. Thus $H^{\mathrm{T}}x = 0$ if and only if $HH^{\mathrm{T}}x = 0$. The same proof applies to show that $\mathscr{N}(H) = \mathscr{N}(H^{\mathrm{T}}H)$. ■

A square matrix is *nonsingular* if its null space consists only of the zero vector. If a square matrix is not nonsingular, it is called *singular*.

(2.12.1) **Exercise:** If the row vectors of H are linearly independent, then the null space of H^{T} consists of the zero vector.

(2.12.2) **Exercise:** Let $h_1, h_2, \ldots, h_n$ be a linearly independent set of vectors. Let G be the $n \times n$ matrix whose $(i-j)$th entry is $h_i^{\mathrm{T}}h_j$ (G is known as a *Grammian*). Show that G is nonsingular. [*Hint:* $G = HH^{\mathrm{T}}$ where H is the matrix whose rows are $h_1^{\mathrm{T}}, h_2^{\mathrm{T}}, \ldots, h_n^{\mathrm{T}}$. Now apply (2.12.1) and (2.12).]

If A is a nonsingular matrix, there is a unique matrix, A^{-1}, which is the left and right inverse of A:

$$A(A^{-1}) = (A^{-1})A = I$$

where I is the identity matrix.

(2.13) Theorem: If H is any matrix and δ is nonzero, then $H^{\mathrm{T}}H + \delta^2 I$ is nonsingular.

Proof: If $(H^{\mathrm{T}}H + \delta^2 I)x = 0$, then $0 = x^{\mathrm{T}}(H^{\mathrm{T}}H + \delta^2 I)x = \|Hx\|^2 + \delta^2 \|x\|^2$ which can only occur if $x = 0$. ■

We close this chapter with a statement of the celebrated diagonalization theorem for symmetric matrices. The proof may be found in Bellman [1].

A (possibly complex) number λ is called an *eigenvalue* of the (square) matrix A if $A - \lambda I$ is singular.

(2.13.1) **Exercise:** If A is real and symmetric, its eigenvalues are real. [*Hint:* If $(A - \lambda I)x = 0$, then $(A - \bar{\lambda} I)\bar{x} = 0$, where $\bar{\lambda}$ and $\bar{x}$ are the complex conjugates of λ and x.]

(2.14) Theorem: If A is real and symmetric with eigenvalues $\lambda_1, \lambda_2, \ldots, \lambda_n$, then there is a matrix T such that $T^{\mathrm{T}} = T^{-1}$ and

$$T^{\mathrm{T}}AT = \operatorname{diag}(\lambda_1, \lambda_2, \ldots, \lambda_n).$$

[The term $\operatorname{diag}(\lambda_1, \ldots, \lambda_n)$ refers to a diagonal matrix with entries $\lambda_1, \ldots, \lambda_n$. If $T^{\mathrm{T}} = T^{-1}$, T is said to be an *orthogonal* matrix.]

(2.14.1) **Exercise:** If T is an orthogonal matrix, the rows of T are mutually orthogonal and have unit length. So too, are the columns.

Chapter III

GEOMETRIC AND ANALYTIC PROPERTIES OF THE MOORE–PENROSE PSEUDOINVERSE

We begin our treatment of the pseudoinverse by characterizing the minimum norm solution to the classical least squares problem:

(3.1) Theorem: Let z be an n-dimensional vector and let H be an $n \times m$ matrix.

(a) There is always a vector, in fact a unique vector $\hat{x}$ of minimum norm, which minimizes

$$\|z - Hx\|^2.$$

(b) $\hat{x}$ is the unique vector in $\mathscr{R}(H^{\mathrm{T}})$ which satisfies the equation

$$Hx = \hat{z}$$

where $\hat{z}$ is the projection of z on $\mathscr{R}(H)$.

Proof: By (2.11) we can write

$$z = \hat{z} + \tilde{z}$$

where $\tilde{z}$ is the projection of z on $\mathscr{N}(H^{\mathrm{T}})$. Since $Hx \in \mathscr{R}(H)$ for every x, it follows that

$$\hat{z} - Hx \in \mathscr{R}(H) \qquad \text{and since} \qquad \tilde{z} \in \mathscr{R}^{\perp}(H), \qquad \tilde{z} \perp \hat{z} - Hx.$$

Therefore

$$\|z-Hx\|^2 = \|\hat{z}-Hx+\tilde{z}\|^2 = \|\hat{z}-Hx\|^2 + \|\tilde{z}\|^2 \geqslant \|\tilde{z}\|^2.$$

This lower bound is attainable since $\hat{z}$, being in the range of H, is the afterimage of some x_0:

$$\hat{z} = Hx_0.$$

Thus, for this x_0, the bound is attained:

$$\|z-Hx_0\|^2 = \|z-\hat{z}\|^2 = \|\tilde{z}\|^2.$$

On the other hand, we just showed that

$$\|z-Hx\|^2 = \|\hat{z}-Hx\|^2 + \|\tilde{z}\|^2$$

so that the lower bound is attained at x^* only if x^* is such that $Hx^* = \hat{z}$. For any such x^*, we can decompose it via (2.11) into two orthogonal vectors:

$$x^* = \hat{x}^* + \tilde{x}^*$$

where

$$\hat{x}^* \in \mathscr{R}(H^{\mathrm{T}}) \qquad \text{and} \qquad \tilde{x}^* \in \mathscr{N}(H).$$

Thus

$$Hx^* = H\hat{x}^* \qquad \text{so that} \qquad \|z-Hx^*\|^2 = \|z-H\hat{x}^*\|^2$$

and

$$\|x^*\|^2 = \|\hat{x}^*\|^2 + \|\tilde{x}^*\|^2 \geqslant \|\hat{x}^*\|^2$$

with strict inequality unless $x^* = \hat{x}^*$ [i.e., unless x^* coincides with its projection on $\mathscr{R}(H^{\mathrm{T}})$—which is to say, unless $x^* \in \mathscr{R}(H^{\mathrm{T}})$ to begin with].

So far, we have shown that x_0 minimizes $\|z-Hx\|^2$ if and only if $Hx_0 = \hat{z}$, and that, among those vectors which minimize $\|z-Hx\|^2$, any vector of minimum norm must lie in the range of H^{T}. To demonstrate the uniqueness of this vector, suppose that $\hat{x}$ and x^* both are in $\mathscr{R}(H^{\mathrm{T}})$ and that

$$H\hat{x} = Hx^* = \hat{z}.$$

Then

$$x^* - \hat{x} \in \mathscr{R}(H^{\mathrm{T}}).$$

But

$$H(x^*-\hat{x}) = 0$$

so that

$$x^* - \hat{x} \in \mathscr{N}(H) = \mathscr{R}^{\perp}(H^{\mathrm{T}}) \quad \text{as well.} \tag{2.10}$$

Thus $x^*-\hat{x}$ is orthogonal to itself, which means that $\|x^*-\hat{x}\|^2 = 0$ (i.e., $x^* = \hat{x}$). ■

(3.1.1) *Comment:* An alternate statement of (3.1) which is equivalent, but perhaps more illuminating is this:

There is always an n-dimensional vector y such that

$$\|z-HH^{\mathrm{T}}y\|^2 = \inf_x \|z-Hx\|^2.$$

If

$$\|z-Hx_0\|^2 = \inf_x \|z-Hx\|^2$$

then $\|x_0\| \geqslant \|H^{\mathrm{T}}y\|$, with strict inequality holding unless $x_0 = H^{\mathrm{T}}y$. y satisfies the equation

$$HH^{\mathrm{T}}y = \hat{z}$$

where $\hat{z}$ is the projection of z on $\mathscr{R}(H)$.

(3.1.2) **Exercise:** $\|z-Hx\|^2$ is minimized by x_0 if and only if $Hx_0 = \hat{z}$, where $\hat{z}$ is the projection of z on $\mathscr{R}(H)$.

The minimal least squares solution alluded to in (3.1) can be characterized as a solution to the so-called *normal equations*:

(3.2) Theorem: Among those vectors x, which minimize $\|z-Hx\|^2$, $\hat{x}$, the one having minimum norm, is the unique vector of the form

$$\hat{x} = H^{\mathrm{T}}y \tag{3.2.1}$$

which satisfies

$$H^{\mathrm{T}}Hx = H^{\mathrm{T}}z. \tag{3.2.2}$$

Comment: The theorem says that $\hat{x}$ can be obtained by finding any vector y_0 which satisfies the equation

$$H^{\mathrm{T}}HH^{\mathrm{T}}y = H^{\mathrm{T}}z$$

and then taking

$$\hat{x} = H^{\mathrm{T}}y_0.$$

Proof: By (2.12), $\mathscr{R}(H^{\mathrm{T}}) = \mathscr{R}(H^{\mathrm{T}}H)$. Since $H^{\mathrm{T}}z$ is in the range of H^{T}, it must therefore be in the range of $H^{\mathrm{T}}H$ and so, must be the afterimage of some x under the transformation $H^{\mathrm{T}}H$. In other words, (3.2.2) always has at least one solution in x. If x is a solution to (3.2.2), so then is $\hat{x}$, the projection of x on $\mathscr{R}(H^{\mathrm{T}})$, since $Hx = H\hat{x}$, (2.11.1). Since $\hat{x} \in \mathscr{R}(H^{\mathrm{T}})$, it is the afterimage of some vector y under H^{T}:

$$\hat{x} = H^{\mathrm{T}}y.$$

So far we have shown that there is at least one solution to (3.2.2) of the form (3.2.1). To show uniqueness, suppose

$$\hat{x}_1 = H^{\mathrm{T}} y_1 \qquad \text{and} \qquad \hat{x}_2 = H^{\mathrm{T}} y_2$$

both satisfy (3.2.2). Then

$$H^{\mathrm{T}} H (H^{\mathrm{T}} y_1 - H^{\mathrm{T}} y_2) = 0$$

so that

$$H^{\mathrm{T}}(y_1 - y_2) \in \mathcal{N}(H^{\mathrm{T}} H) = \mathcal{N}(H) \tag{2.12}$$

which implies that

$$H H^{\mathrm{T}}(y_1 - y_2) = 0.$$

Thus

$$(y_1 - y_2) \in \mathcal{N}(H H^{\mathrm{T}}) = \mathcal{N}(H^{\mathrm{T}}) \tag{2.12}$$

and so

$$\hat{x}_1 = H^{\mathrm{T}} y_1 = H^{\mathrm{T}} y_2 = \hat{x}_2.$$

Thus, there is *exactly one* solution to (3.2.2) of the form (3.2.1). If we can show that this solution also satisfies the equation

$$Hx = \hat{z}$$

where $\hat{z}$ is the projection of z on $\mathscr{R}(H)$ then, by virtue of (3.1b) we will be done.

But, in (2.11.1) we showed that

(3.2.3) $$H^{\mathrm{T}} z = H^{\mathrm{T}} \hat{z}.$$

In Theorem (3.1), we showed that there is a unique solution in $\mathscr{R}(H^{\mathrm{T}})$ to the equation

(3.2.4) $$Hx = \hat{z}.$$

This (unique) solution therefore satisfies the equation

$$H^{\mathrm{T}} H x = H^{\mathrm{T}} \hat{z}$$

as well. Since $H^{\mathrm{T}} z = H^{\mathrm{T}} \hat{z}$, (3.2.3), we see that the (unique) solution to (3.2.4) which lies in $\mathscr{R}(H^{\mathrm{T}})$ must coincide with $\hat{x}$, the unique solution to (3.2.2) which lies in $\mathscr{R}(H^{\mathrm{T}})$. In summary, the vector $\hat{x}$ alluded to in the statement of (3.2) coincides exactly with the vector $\hat{x}$ alluded to in (3.1). ■

We are now in a position to exhibit an explicit representation for the minimum norm solution to a least squares problem. A preliminary lemma is needed which, for the sake of continuity, is stated here and proven later:

(3.3) Lemma: For any real symmetric matrix A,

$$P_A = \lim_{\delta\to 0}(A+\delta I)^{-1}A = \lim_{\delta\to 0}A(A+\delta I)^{-1}$$

always exists. For any vector z,

$$\hat{z} = P_A z$$

is the projection of z on $\mathscr{R}(A)$.

(3.4) Theorem: For any $n\times m$ matrix H,

(3.4.1) $$H^+ = \lim_{\delta\to 0}(H^TH+\delta^2 I)^{-1}H^T$$

(3.4.2) $$= \lim_{\delta\to 0}H^T(HH^T+\delta^2 I)^{-1}$$

always exists. For any n-vector z,

$$\hat{x} = H^+ z$$

is the vector of minimum norm among those which minimize

$$\|z-Hx\|^2.$$

Comment: Henceforth, we use the symbol I to represent the identity matrix, whose dimensionality is to be understood from the context. For example, in the expression H^TH+I, we refer to the $m\times m$ identity, whereas in HH^T+I, we refer to the $n\times n$ identity.

Proof: Since

$$(H^THH^T+\delta^2 H^T) = H^T(HH^T+\delta^2 I) = (H^TH+\delta^2 I)H^T$$

and since $(HH^T+\delta^2 I)$ and $(H^TH+\delta^2 I)$ have inverses when $\delta^2>0$, (2.13), it is clear that the right sides of (3.4.1) and (3.4.2) are equal if either exists.

Let z be a given n-dimensional vector, and decompose z into its projections on $\mathscr{R}(H)$ and $\mathscr{N}(H^T)$ according to (2.11);

$$z = \hat{z} + \tilde{z}.$$

Since

$$H^Tz = H^T\hat{z} \qquad (2.11.1)$$

and since $\hat{z}\in\mathscr{R}(H)$ must be the afterimage of some vector x_0 under H, we see that

(3.4.3) $$(H^TH+\delta^2 I)^{-1}H^Tz = (H^TH+\delta^2 I)^{-1}H^T\hat{z}$$

$$= (H^TH+\delta^2 I)^{-1}H^THx_0.$$

The limit of the last expression always exists and coincides with $\hat{x}_0$, the projection of x_0 on $\mathscr{R}(H^{\mathrm{T}}H)$, by virtue of (3.3). Since $\mathscr{R}(H^{\mathrm{T}}) = \mathscr{R}(H^{\mathrm{T}}H)$, (2.12), and since

$$\begin{aligned} \hat{z} &= Hx_0 \\ &= H\hat{x}_0 \end{aligned} \tag{2.11.1}$$

we conclude that

$$\hat{x}_0 \equiv \lim_{\delta \to 0} (H^{\mathrm{T}}H + \delta^2 I)^{-1} H^{\mathrm{T}} z$$

always exists, is an element of $\mathscr{R}(H^{\mathrm{T}})$, and satisfies the relation

$$H\hat{x}_0 = \hat{z}$$

where $\hat{z}$ is the projection of z on $\mathscr{R}(H)$. The desired conclusion follows directly from (3.1). ■

(3.5) Corollary: For any vector z, HH^+z is the projection of z on $\mathscr{R}(H)$ and $(I-HH^+)z$ is the projection of z on $\mathscr{N}(H^{\mathrm{T}})$. For any vector x, H^+Hx is the projection of x on $\mathscr{R}(H^{\mathrm{T}})$ and $(I-H^+H)x$ is the projection of x on $\mathscr{N}(H)$.

Proof: By (3.4.2), $HH^+ = \lim_{\delta\to 0} HH^{\mathrm{T}}(HH^{\mathrm{T}}+\delta^2 I)^{-1}$ and by (3.4.1), $H^+H = \lim_{\delta \to 0}(H^{\mathrm{T}}H+\delta^2 I)^{-1}H^{\mathrm{T}}H$. (3.3) tells us that HH^+z is therefore the projection of z on $\mathscr{R}(HH^{\mathrm{T}})$, which coincides with the projection of z on $\mathscr{R}(H)$, (2.12). Similarly, (3.3) and (2.12) imply that H^+Hx is the projection of x on $\mathscr{R}(H^{\mathrm{T}}H) = \mathscr{R}(H^{\mathrm{T}})$.

Since $z-\hat{z}$ is the projection of z on $\mathscr{N}(H^{\mathrm{T}})$ if $\hat{z}$ is the projection of z on $\mathscr{R}(H)$, (2.11), it follows that

$$z - HH^+z \qquad \text{is the projection of } z \text{ on } \mathscr{N}(H^{\mathrm{T}}).$$

By the same token,

$$(I-H^+H)x \qquad \text{is the projection of } x \text{ on } \mathscr{N}(H). \quad ■$$

The matrix H^+, which we explicitly define in (3.4), is the so-called "Moore–Penrose generalized inverse for H." We will usually refer to it more familiarly as "the pseudoinverse of H."

Corollary (3.5) is tremendously important and should be noted carefully. It expresses the four most fundamental projection operators in terms of pseudoinverses. The results of (3.4) are attributed (via a slightly different method of proof) to den Broeder and Charnes [1].

(3.5.1) **Exercise:** $H^+ = H^{-1}$ if H is square and nonsingular.

(3.5.2) **Exercise:** $H^+ = H^T(HH^T)^{-1}$ if the rows of H are linearly independent. [*Hint:* Apply (2.12.2) to show (HH^T) has an inverse. Then use (3.4.2).]

(3.5.3) **Exercise:** $H^+ = (H^TH)^{-1}H^T$ if the columns of H are linearly independent.

Before proceeding to the light task of milking these results for all they are worth, we pause to furnish the proof for Lemma (3.3), as promised:

Proof of (3.3): If A is any symmetric matrix and δ_0 is a nonzero scalar whose magnitude is less than the magnitude of A's smallest nonzero eigenvalue, then for any δ with

$$0 < |\delta| < |\delta_0|$$

$(A+\delta I)$ is nonsingular and hence, for all such δ's, $(A+\delta I)^{-1}$ exists.

If z is any vector, we write

$$z = \hat{z} + \tilde{z}$$

where

$$\hat{z} \in \mathscr{R}(A) \qquad \tilde{z} \in \mathscr{N}(A) \tag{2.11}$$

and

$$Az = A\hat{z}. \tag{2.11.1}$$

Since $\hat{z} \in \mathscr{R}(A)$, we can write $\hat{z} = Ax_0$ for some x_0 and so

$$\begin{aligned}(A+\delta I)^{-1}Az &= (A+\delta I)^{-1}A\hat{z}\\ &= (A+\delta I)^{-1}A(Ax_0).\end{aligned}$$

From (2.14), the diagonalization theorem, we conclude that

$$A = TDT^{T}$$

where

$$D = \mathrm{diag}(\lambda_1, \lambda_2, \ldots, \lambda_n)$$

is the diagonal matrix of A's eigenvalues and T is an orthogonal matrix:

$$T^T = T^{-1}.$$

Thus

$$(A+\delta I)^{-1}Az = (A+\delta I)^{-1}A^2x_0 = T(D+\delta I)^{-1}D^2T^Tx_0.$$

Element-by-element, it is plain to see that

$$\lim_{\delta\to 0}(D+\delta I)^{-1}D^2 = D$$

so that

$$\lim_{\delta\to 0}(A+\delta I)^{-1}Az = TDT^{\mathrm{T}}x_0 = Ax_0 = \hat{z}$$

the projection of z on $\mathscr{R}(A)$.

The same argument works for $\lim_{\delta\to 0} A(A+\delta I)^{-1}z$. ■

In (3.5.1)–(3.5.3), formulas for H^+ in terms of inverses were given for the case where the rows and/or the columns of H are linearly independent. There are cases though where these conditions need not obtain:

$$H = \begin{pmatrix} 1 & 1 \\ 1 & 1 \end{pmatrix}.$$

In such cases, H^+ does not have a simple formula in terms of inverses. However, a better understanding of H^+ can be had by treating, in turn, the following cases: H a 1×1 matrix, H diagonal, H symmetric, H rectangular:

If H is a 1×1 (scalar) matrix, then

$$H^+ = \lim_{\delta^2\to 0}(H^2+\delta^2 I)H = \begin{cases} 0 & \text{if } H = 0 \\ 1/H & \text{if } H \neq 0. \end{cases}$$

If H is diagonal:

$$H = \operatorname{diag}(\lambda_1, \lambda_2, \ldots, \lambda_m)$$

then

$$H^+ = \operatorname{diag}(\lambda_1{}^+, \lambda_2{}^+, \ldots, \lambda_m{}^+)$$

where

$$\lambda_j{}^+ = \begin{cases} 0 & \text{if } \lambda_j = 0 \\ 1/\lambda_j & \text{if } \lambda_j \neq 0. \end{cases}$$

In terms of least squares, this makes sense, for if

$$z = \begin{pmatrix} \zeta_1 \\ \vdots \\ \zeta_n \end{pmatrix} \qquad H = \operatorname{diag}(\lambda_1, \lambda_2, \ldots, \lambda_n)$$

and

$$x = \begin{pmatrix} \xi_1 \\ \vdots \\ \xi_n \end{pmatrix}$$

is to be chosen to minimize

$$\|z-Hx\|^2 = \sum_{=1}^{n}(\zeta_j-\lambda_j\xi_j)^2$$

it is clear that the choice

$$\xi_j^* = \begin{cases} \zeta_j/\lambda_j & \text{if} \quad \lambda_j \neq 0 \\ \text{arbitrary} & \text{if} \quad \lambda_j = 0 \end{cases}$$

will minimize the sum of squares, and that

$$\|x^*\|^2 = \sum_{j=1}^{n} \xi_j^{*2}$$

is made smallest when

$$\xi_j^* = 0 \qquad \text{if} \quad \lambda_j = 0.$$

Thus, the minimum norm solution for the case of a diagonal H is

$$\hat{x} = \begin{pmatrix} \hat{\xi}_1 \\ \vdots \\ \hat{\xi}_n \end{pmatrix}$$

where

$$\hat{\xi}_j = \lambda_j^+ \zeta_j; \qquad \text{i.e.,} \qquad \hat{x} = H^+ z.$$

(3.6) The Special Case of Symmetric Matrices

If H is a symmetric $m \times m$ matrix, the diagonalization theorem allows us to write

$$H = TDT^{\mathrm{T}}$$

where T is an orthogonal matrix and D is diagonal.

By (3.4),

$$\begin{aligned} H^+ &= \lim_{\delta \to 0} T(D^2 + \delta^2 I)^{-1} DT^{\mathrm{T}} \\ &= T[\lim_{\delta \to 0} (D^2 + \delta^2 I)^{-1} D]\, T^{\mathrm{T}} \\ &= TD^+ T^{\mathrm{T}}. \end{aligned}$$

Thus, the pseudoinverse for a symmetric matrix is obtained by pseudo-inverting the diagonal matrix of its eigenvalues. Since H is nonsingular if and only if its eigenvalues are nonzero (in which case $D^+ = D^{-1}$), we see that

$$H^+ = TD^{-1}T^{\mathrm{T}}$$

if H is symmetric and nonsingular: Since $TT^{\mathrm{T}} = T^{\mathrm{T}}T = I$, it is easy to see that $HH^+ = H^+H = I$ in this case, so that $H^+ = H^{-1}$.

The last result can be expressed in a different notation, and gives rise to various so-called spectral representation theorems [cf. (3.15)]: If the column vectors of T are denoted by t_i $(i = 1, \ldots, m)$, so that we can write T as a partitioned matrix,

$$T = (t_1 \vdots t_2 \vdots \cdots \vdots t_m)$$

the diagonalization theorem states that

(3.6.1) $$H = TDT^{\mathrm{T}} \equiv \sum_{=1}^{m} \lambda_j t_j t_j^{\mathrm{T}}.$$

Furthermore, since

$$T^{\mathrm{T}}T = I$$

this tells us that

$$t_i^{\mathrm{T}} t_j = \begin{cases} 1 & \text{if } i = j \\ 0 & \text{otherwise} \end{cases}$$

so that the columns of T are *orthonormal* (mutually orthogonal with unit length). Furthermore,

$$HT = TDT^{\mathrm{T}}T = TD$$

which can be read column by column as

$$Ht_j = \lambda_j t_j \qquad j = 1, 2, \ldots, m.$$

Thus, each t_j is an *eigenvector* of H associated with the eigenvalue λ_j. If the λ_j's are not all distinct, the t_j's are mutually orthogonal nonetheless.

The fact that

$$H^+ = TD^+T^{\mathrm{T}}$$

can be expressed as

(3.6.2) $$H^+ = \sum_{j=1}^{m} \lambda_j^+ t_j t_j^{\mathrm{T}}.$$

(3.7) Exercises

(3.7.1) Let H be an $m \times m$ symmetric matrix and suppose the nonzero eigenvalues of H are $\lambda_1, \lambda_2, \ldots, \lambda_k$ $(k \leqslant m)$.

(a) Show that there is a representation for H of the form

$$H = TDT^{\mathrm{T}}$$

where

$$D = \operatorname{diag}(\lambda_1, \lambda_2, \ldots, \lambda_k, \overbrace{0, \ldots, 0}^{m-k \text{ zeros}})$$

and T is an orthogonal matrix. (*Hint:* If D is not arranged properly, use a permutation matrix, which is always orthogonal, to do so.)

(b) If the columns of T are denoted by $t_1, t_2, \ldots, t_m$, show that

$$\mathscr{R}(H) = \mathscr{L}(t_1, \ldots, t_k) \quad \text{and} \quad \mathscr{N}(H) = \mathscr{L}(t_{k+1}, \ldots, t_m).$$

(3.7.2) Without appealing to the diagonalization theorem, show directly that if H is symmetric and if $Hx = \lambda_1 x$ and $Hy = \lambda_2 y$ then $x \perp y$ if $\lambda_1 \neq \lambda_2$.

The representation (3.6.2) is particularly interesting since it clearly shows the radically discontinuous nature of pseudoinversion. Two matrices may be very close to each other element by element. However, if their ranks differ (e.g., if one is singular, while the other is nonsingular) their pseudoinverses usually differ greatly. For example, the diagonal matrices

$$D_1 = \begin{pmatrix} 4 & 0 \\ 0 & 0 \end{pmatrix} \quad \text{and} \quad D_2 = \begin{pmatrix} 4 & 0 \\ 0 & 10^{-10} \end{pmatrix}$$

are close to each other, but

$$D_1{}^+ = \begin{pmatrix} \frac{1}{4} & 0 \\ 0 & 0 \end{pmatrix} \quad \text{and} \quad D_2{}^+ = \begin{pmatrix} \frac{1}{4} & 0 \\ 0 & 10^{10} \end{pmatrix}$$

differ greatly. In terms of (3.6.2), it is easy to understand why, since the transformation

$$\lambda^+ = \begin{cases} 1/\lambda & \text{if } \lambda \neq 0 \\ 0 & \text{if } \lambda = 0 \end{cases}$$

exhibits an infinite discontinuity at $\lambda = 0$. This characteristic induces serious computational difficulties which we will discuss at the appropriate time.

In (3.8), we will see that the pseudoinverse of an arbitrary rectangular matrix is expressible in terms of the pseudoinverse of symmetric matrices:

$$\begin{aligned} H^+ &= (H^T H)^+ H^T \\ &= H^T (H H^T)^+ \end{aligned}$$

so that one can, in theory, diagonalize symmetric matrices (for which well-known algorithms are available) and proceed directly to pseudoinversion from there. However, there are other, less tedious methods for computing H^+ (Chapter V). The problem of round-off is quite serious, as the reader may have already come to appreciate.

(3.7.3) Let H be an $n \times m$ matrix and let $x_1, x_2, \ldots, x_r$ $r \leqslant m$ be an orthonormal set of n-dimensional vectors such that

$$\mathscr{R}(H) = \mathscr{L}(x_1, x_2, \ldots, x_r).$$

Then

$$HH^+ = \sum_{j=1}^{r} x_j x_j^{\mathrm{T}}.$$

[*Hint:* Use (3.5) and (2.7).]

A symmetric matrix P is called a *projection matrix* if it is idempotent (i.e., $P^2 = P$).

(3.7.4) The eigenvalues of a projection matrix are either zero or unity.

(3.7.5) If P is a projection matrix, $P^+ = P$ and Px is the projection of x on $\mathscr{R}(P)$. If $x \in \mathscr{R}(P)$, then $Px = x$.

(3.7.6) HH^+, H^+H, $I - HH^+$, and $I - H^+H$ are projection matrices.

(3.7.7) If $P_1, P_2, \ldots, P_n$ are projection matrices having the property that $P_i P_j = 0$ if $i \neq j$ and if P is another projection such that

$$\mathscr{R}(P) = \mathscr{R}\left(\sum_{j=1}^{n} P_j\right) \qquad \text{then} \quad P = \sum_{j=1}^{n} P_j.$$

[*Hint:* First show that $Q \equiv \sum_{j=1}^{n} P_j$ is a projection matrix. Then $Q^+ = Q$ and so $QQ^+x = Qx$ which is the projection of x on $\mathscr{R}(Q)$. On the other hand, $PP^+x = Px$ is the projection of x on the range of P. Since $\mathscr{R}(P) = \mathscr{R}(Q)$, we have $Px = Qx$ for all x.]

(3.7.8) Let $h_1, h_2, \ldots, h_n$ be a set of vectors and let H be the matrix whose column vectors are $h_1, h_2, \ldots, h_n$. Then for any x, HH^+x is the projection of x on $\mathscr{L}(h_1, h_2, \ldots, h_n)$.

(3.7.9) If H is any matrix, then $Hx = 0$ if and only if $x = (I - H^+H)y$ for some y.

(3.7.10) If H is any matrix, then $z \in \mathscr{R}(H)$ if and only if $z = HH^+u$ for some u.

We now turn our attention to an exploration of the most important properties of H^+.

(3.8) Theorem: For any matrix H,

(3.8.1) $$H^+ = (H^{\mathrm{T}}H)^+ H^{\mathrm{T}}$$

(3.8.2) $$(H^{\mathrm{T}})^+ = (H^+)^{\mathrm{T}}$$

(3.8.3) $$H^+ = H^{\mathrm{T}}(HH^{\mathrm{T}})^+.$$

Proof

$$(H^{\mathrm{T}}H)^+ H^{\mathrm{T}} = \{\lim_{\delta\to 0}[(H^{\mathrm{T}}H)^2+\delta^2 I]^{-1}(H^{\mathrm{T}}H)\} H^{\mathrm{T}}$$

and

$$H^+ = \lim_{\delta\to 0}[H^{\mathrm{T}}H+\delta^2 I]^{-1} H^{\mathrm{T}}. \qquad (3.4)$$

By (2.11), any vector z can be written as

$$z = Hx_0 + \tilde{z} \qquad \text{for some} \quad x_0$$

where

$$H^{\mathrm{T}}\tilde{z} = 0.$$

Thus

(3.8.4) $$(H^{\mathrm{T}}H)^+ H^{\mathrm{T}}z = \lim_{\delta\to 0}[(H^{\mathrm{T}}H)^2+\delta^2 I]^{-1}(H^{\mathrm{T}}H)^2 x_0$$

and

(3.8.5) $$H^+z = \lim_{\delta\to 0}[H^{\mathrm{T}}H+\delta^2 I]^{-1} H^{\mathrm{T}}Hx_0.$$

Using the diagonalization theorem (2.14), we write

$$H^{\mathrm{T}}H = TDT^{\mathrm{T}}$$

where D is diagonal and T is orthogonal. Equation (3.8.4) takes the form

$$(H^{\mathrm{T}}H)^+ H^{\mathrm{T}}z = T\{\lim_{\delta\to 0}[D^2+\delta^2 I]^{-1}D^2\}\, T^{\mathrm{T}}x_0$$

while (3.8.5) takes the form

$$H^{\mathrm{T}}z = T\{\lim_{\delta\to 0}[D+\delta^2 I]^{-1}D\}\, T^{\mathrm{T}}x_0.$$

For diagonal D's, the matrices in braces are the same, which proves (3.8.1).

To prove (3.8.2), notice that $(HH^{\mathrm{T}}+\delta^2 I)^{-1}$ is symmetric so by (3.4.1),

$$\begin{aligned}(H^{\mathrm{T}})^+ &= \lim_{\delta\to 0}(HH^{\mathrm{T}}+\delta^2 I)^{-1}H \\ &= \lim_{\delta\to 0}[H^{\mathrm{T}}(HH^{\mathrm{T}}+\delta^2 I)^{-1}]^{\mathrm{T}} \\ &= (H^+)^{\mathrm{T}}. \qquad (3.4.2)\end{aligned}$$

To prove (3.8.3), we use (3.8.2) to establish that

(3.8.6) $$[(H^{\mathrm{T}})^+]^{\mathrm{T}} = H^+$$

and (3.8.1) to establish that

(3.8.7) $$(H^{\mathrm{T}})^{+} = (HH^{\mathrm{T}})^{+} H.$$

Since $(HH^{\mathrm{T}})^{+}$ is symmetric, (3.8.3) follows from (3.8.6) after taking transposes in (3.8.7). ■

In his paper of 1955, which was probably responsible for the rebirth of interest in the topic of generalized inverses, Penrose [1], characterized the pseudoinverse as the (unique) solution to a set of matrix equations. The pseudoinverse H^{+} that we have been investigating satisfies the Penrose conditions:

(3.9) Theorem: For any matrix H, $B = H^{+}$ if and only if

(3.9.1) HB and BH are symmetric

(3.9.2) $$HBH = H$$

(3.9.3) $$BHB = B.$$

Proof of necessity

$$HH^{+} = \lim_{\delta\to 0} HH^{\mathrm{T}}(HH^{\mathrm{T}}+\delta^2 I)^{-1}$$

and

$$H^{+}H = \lim_{\delta\to 0}(H^{\mathrm{T}}H+\delta^2 I)^{-1} H^{\mathrm{T}}H. \qquad (3.4)$$

Both are symmetric. This shows that H^{+} satisfies (3.9.1). By (3.5), HH^{+} is the projection on $\mathscr{R}(H)$. Since $Hx \in \mathscr{R}(H)$ for all x, (3.7.5) assures that $(HH^{+})(Hx) = Hx$. This shows that H^{+} satisfies (3.9.2). By (3.8.1),

(3.9.4) $$H^{+}H = (H^{\mathrm{T}}H)^{+}(H^{\mathrm{T}}H).$$

By (3.8.1) and (3.9.2)

$$H^{+} = (H^{\mathrm{T}}H)^{+}H^{\mathrm{T}} = (H^{\mathrm{T}}H)^{+}[H(H^{+}H)]^{\mathrm{T}}$$
$$= (H^{\mathrm{T}}H)^{+}H^{\mathrm{T}}(HH^{+})^{\mathrm{T}}.$$

Since HH^{+} is symmetric, (3.7.6),

$$H^{+} = (H^{\mathrm{T}}H)^{+}H^{\mathrm{T}}(HH^{+}) = (H^{\mathrm{T}}H)^{+}(H^{\mathrm{T}}H)H^{+}$$

and by (3.9.4), the last is equal to

$$(H^{+}H)H^{+}.$$

This establishes (3.9.3).

Proof of sufficiency: Suppose B satisfies (3.9.1)–(3.9.3):
Since

$$BH = (BH)^{\mathrm{T}} \quad \text{and} \quad H = HBH$$

$$H = HBH = HH^{\mathrm{T}}B^{\mathrm{T}}.$$

Since

$$HH^+H = H$$

$$H^+H = H^+(HH^{\mathrm{T}}B^{\mathrm{T}}) = [H(H^+H)]^{\mathrm{T}}B^{\mathrm{T}}$$

and so

(3.9.5) $$H^+H = H^{\mathrm{T}}B^{\mathrm{T}} = BH.$$

Since $B = BHB$ and since HB is symmetric,

(3.9.6) $$B^{\mathrm{T}} = HBB^{\mathrm{T}}.$$

Premultiplying (3.9.6) by HH^+, we find that

$$HH^+B^{\mathrm{T}} = HH^+HBB^{\mathrm{T}} = HBB^{\mathrm{T}} \qquad (3.9.2)$$

and by (3.9.6) the last is equal to B^{T}. Thus

(3.9.7) $$B^{\mathrm{T}} = (HH^+)B^{\mathrm{T}}.$$

Taking transposes in (3.9.7), we find that

$$B = B(HH^+)^{\mathrm{T}} = (BH)H^+$$

and by (3.9.5) we finally conclude that

$$B = H^+HH^+.$$

Since $H^+HH^+ = H^+$, we see that

$$B = H^+. \quad ■$$

The Penrose characterization for pseudoinverses is extremely useful as a method for proving identities. For instance, if one thinks that a certain expression coincides with the pseudoinverse of a certain matrix H, a handy way of deciding is to run the expression through conditions (3.9.1)–(3.9.3) and observe whether or not they hold.

(3.10) Exercise: If A and B are nonsingular, it is well known that $(AB)^{-1} = B^{-1}A^{-1}$. Use (3.9) to show that it is not generally true that $(AB)^+ = B^+A^+$. Where do the conditions break down? Exhibit a counterexample. [See (4.10)–(4.16) for a detailed study of this problem.]

(3.11) Exercise: Prove the following:

(3.11.1) $(H^+)^+ = H.$

(3.11.2) $(H^{\mathrm{T}}H)^+ = H^+(H^{\mathrm{T}})^+$ and $(HH^{\mathrm{T}})^+ = (H^{\mathrm{T}})^+H^+.$

(3.11.3) If A is symmetric and $\alpha > 0$, then $(A^\alpha)^+ = (A^+)^\alpha$ and $A^\alpha(A^\alpha)^+ = (A^\alpha)^+A^\alpha = AA^+.$

(3.11.4) $(H^{\mathrm{T}}H)^+ = H^+(HH^{\mathrm{T}})^+H = H^{\mathrm{T}}(HH^{\mathrm{T}})^+(H^{\mathrm{T}})^+.$

(3.11.5) $\mathscr{R}(H^+) = \mathscr{R}(H^+H) = \mathscr{R}(H^{\mathrm{T}})$; $\mathscr{N}(H) = \mathscr{N}(H^+H) = \mathscr{N}[(H^{\mathrm{T}}H)^+].$

(3.11.6) If A is symmetric, $AA^+ = A^+A.$

(3.11.7) $HH^+ = (HH^{\mathrm{T}})(HH^{\mathrm{T}})^+ = (HH^{\mathrm{T}})^+(HH^{\mathrm{T}})$ and $H^+H = (H^{\mathrm{T}}H)(H^{\mathrm{T}}H)^+ = (H^{\mathrm{T}}H)^+(H^{\mathrm{T}}H).$

(3.11.8) If A is symmetric and $\alpha > 0$, $A^+A^\alpha = A^\alpha A^+.$

(3.11.9) If H is a nonzero $n \times 1$ matrix (a column vector) $H^+ = H^{\mathrm{T}}/H^{\mathrm{T}}H$ and $HH^+ = HH^{\mathrm{T}}/\|H\|^2.$

The properties of pseudoinverses and projections which we have developed can be readily applied to the theory of least squares subject to constraints. But first, we summarize the general results for unconstrained least squares and the related theory of linear equations, in the language of pseudoinverses:

(3.12) Theorem: (a) x_0 minimizes

(3.12.1)
$$\|z - Hx\|^2$$

if and only if x_0 is of the form

(3.12.2)
$$x_0 = H^+z + (I - H^+H)y$$

for some y.

(b) The value of x which minimizes (3.12.1) is unique if and only if $H^+H = I$. The last is true if and only if zero is the only null vector of H.

(c) The equation

(3.12.3)
$$Hx = z$$

has a solution if and only if

$$HH^+z = z.$$

The last is true if and only if $z \in \mathscr{R}(H)$. x_0 is a solution to (3.12.3) if and only if it is of the form (3.12.2). Equation (3.12.3) has a unique solution ($= H^+z$) if and only if $HH^+z = z$ and $H^+H = I$.

Proof: (a) $\|z-Hx\|^2$ is minimized by x_0 if and only if $Hx_0 = \hat{z}$, where $\hat{z}$ is the projection of z on $\mathscr{R}(H)$, (3.12). By (3.4), H^+z minimizes $\|z-Hx\|^2$ so that $Hx_0 = H(H^+z)$. This means that $x_0 - H^+z$ is a null vector of H if x_0 minimizes (3.12.1). The last is true if and only if

$$x_0 - H^+z = (I-H^+H)y \qquad \text{for some} \quad y. \tag{3.7.9}$$

Conversely, if x_0 has the form (3.12.2), then $Hx_0 = H(H^+z)$ since $H(I-H^+H) = 0$, (3.9.2). This proves part (a).

(b) The value of x which minimizes (3.12.1) is unique if and only if $(I-H^+H)y$ vanishes for all y. This can only happen if the projection of all vectors on $\mathscr{N}(H)$ is zero, which means that $\mathscr{N}(H)$ consists only of the zero vector. $(I-H^+H)y = 0$ for all y, by the way, if and only if $H^+H = I$. This proves (b).

(c) Equation (3.12.3) has a solution if and only if z is the afterimage of some x under H. This is, by definition, the same as saying $z \in \mathscr{R}(H)$. By virtue of (3.7.10), the last holds true if and only if

$$z = HH^+u$$

for some u. Since HH^+ is a projection, (3.7.6), it follows that

$$HH^+z = (HH^+)^2u = HH^+u = z.$$

When (3.12.3) has a solution x_0, this solution must minimize $\|z-Hx\|^2$ (the minimal value in this case being zero) and so x_0 must be of the form (3.12.2). The solution is unique if and only if $H^+H = I$ [part (b)] and these conclusions collectively serve to establish (c). ■

(3.12.4) **Corollary:** Let G be a rectangular matrix and suppose u is a vector in $\mathscr{R}(G)$. Then

(a) $\mathscr{S} = \{x: Gx = u\}$ is nonempty and x_0 minimizes $\|z-Hx\|^2$ over $\mathscr{S}$ if and only if

$$x_0 = G^+u + \bar{H}^+\bar{z} + (I-G^+G)(I-\bar{H}^+\bar{H})y$$

for some y, where

$$\bar{z} = z - HG^+u \qquad \text{and} \qquad \bar{H} = H(I-G^+G).$$

(b) The vector of minimum norm among those which minimize $\|z-Hx\|^2$ over $\mathscr{S}$ is

$$G^+u + \bar{H}^+\bar{z}.$$

Proof: (a) If $u \in \mathscr{R}(G)$ then by (3.12), $\mathscr{S}$ is nonempty and

$$\mathscr{S} = \{x: x = G^+u + (I-G^+G)v \quad \text{for some} \quad v\}.$$

Therefore

$$\min_{x \in \mathscr{S}} \|z - Hx\| = \min_{v} \|\bar{z} - \bar{H}v\|.$$

The latter minimum occurs at v_0 if and only if

$$v_0 = \bar{H}^+ \bar{z} + (I - \bar{H}^+ \bar{H})y \qquad \text{for some} \quad y \tag{3.12a}$$

so that x_0 minimizes $\|z - Hx\|$ over $\mathscr{S}$ if and only if

$$x_0 = G^+ u + (I - G^+ G)[\bar{H}^+ \bar{z} + (I - \bar{H}^+ \bar{H})y] \qquad \text{for some} \quad y.$$

Since

$$\bar{H}^+ = \bar{H}^{\mathrm{T}}(\bar{H}\bar{H}^{\mathrm{T}})^+ \tag{3.8.3}$$

and since

$$(I - G^+ G)^2 = (I - G^+ G) = (I - G^+ G)^{\mathrm{T}} \tag{3.7.6}$$

it follows that

(3.12.4.1) $$(I - G^+ G)\bar{H}^+ = (I - G^+ G)^2 H^{\mathrm{T}}(\bar{H}\bar{H}^{\mathrm{T}})^+ = \bar{H}^+$$

and so it is clear that any value of x which minimizes $\|z - Hx\|$ over $\mathscr{S}$ is of the form

$$x_0 = G^+ u + \bar{H}^+ \bar{z} + (I - G^+ G)(I - \bar{H}^+ \bar{H})y$$

for some y.

(b) $[(I - G^+ G)(I - \bar{H}^+ \bar{H})y]^{\mathrm{T}} G^+ u = y^{\mathrm{T}}(I - \bar{H}^+ \bar{H})(I - G^+ G)G^+ u$ since $(I - G^+ G)$ and $I - \bar{H}^+ \bar{H}$ are symmetric. The last is zero since $(I - G^+ G)G^+ = G^+ - G^+ G G^+ = 0$. Thus

(3.12.4.2) $$(I - G^+ G)(I - \bar{H}^+ \bar{H})y \perp G^+ u.$$

On the other hand,

$$\begin{aligned}[(I - G^+ G)(I - \bar{H}^+ \bar{H})y]^{\mathrm{T}} \bar{H}^+ \bar{z} &= y^{\mathrm{T}}(I - \bar{H}^+ \bar{H})(I - G^+ G)\bar{H}^+ \bar{z} \\ &= y^{\mathrm{T}}(I - \bar{H}^+ \bar{H})\bar{H}^+ \bar{z}. \qquad (3.12.4.1)\end{aligned}$$

Since $\bar{H}^+ - \bar{H}^+ \bar{H}\bar{H}^+ = 0$, we see that $(I - G^+ G)(I - \bar{H}^+ \bar{H})y \perp \bar{H}^+ \bar{z}$ as well, so that if x_0 minimizes $\|z - Hx\|^2$ over $\mathscr{S}$, then

$$\begin{aligned}\|x_0\|^2 &= \|G^+ u + \bar{H}^+ \bar{z}\|^2 + \|(I - G^+ G)(I - \bar{H}^+ \bar{H})y\|^2 \\ &\geqslant \|G^+ u + \bar{H}^+ \bar{z}\|^2\end{aligned}$$

with strict inequality holding unless

$$x_0 = G^+ u + \bar{H}^+ \bar{z}. \qquad \blacksquare$$

(3.12.5) **Exercise:** (a) The equation $Hx = z$ has a solution for all z if the rows of H are linearly independent.

(b) If the equation $Hx = z$ has a solution, the solution is unique if and only if the columns of H are linearly independent.

(3.12.6) **Exercise:** Let H be an $n \times m$ matrix with linearly independent columns. For any $k \times m$ matrix G, let $\bar{H} = H(I-G^+G)$. Show that $(I-G^+G)(I-\bar{H}^+\bar{H}) = 0$. [*Hint:* If $w = (I-G^+G)(I-\bar{H}^+\bar{H})v$, then $Hw = 0$. Apply (2.12.1).]

Comment: If the columns of H are linearly independent and u is in the range of G, then (3.12.6) and (3.12.4) together, imply that there is a unique vector which minimizes $\|z-Hx\|^2$ over $\mathscr{S}$. In general, though, the minimizing vector is not unique. However, the vector, $\hat{x}_0 = G^+u+\bar{H}^+\bar{z}$, has minimum norm among those which minimize $\|z-Hx\|^2$ over $\mathscr{S}$. The vector $\bar{x} = \bar{H}^+\bar{z}$ is the vector of minimum norm among those which minimize $\|\bar{z}-\bar{H}x\|^2$ subject to no constraints. $\bar{x}$ and $\hat{x}_0$ differ by G^+u, the minimum norm vector in $\mathscr{S}$.

(3.12.7) **Exercise:** [Refer to (3.12.4).] If $\mathscr{S}$ is empty, then $x^* = G^+u+\bar{H}^+\bar{z}$ is the vector of minimum norm which minimizes $\|z-Hx\|$ over $\mathscr{S}^*$ where $\mathscr{S}^* = \{x\colon \|Gx-u\|^2$ is minimized$\}$. [*Hint:* $\mathscr{S}^* = \{x\colon x = G^+u-(I-G^+G)v$ for some $v.\}$ The proof of (3.12.4) carries over word for word.]

(3.12.8) **Application to Linear Programming** (Ben-Israel and Charnes [2], also Ben-Israel *et al.* [1])

Let a, b, and c be specified n-vectors and let A be a given $m \times n$ matrix. Consider the problem of minimizing c^Tx with respect to x, subject to the constraints

$$a \leqslant Ax \leqslant b$$

(where the vector inequality is to be interpreted componentwise).

Following the conventional terminology, the problem is said to be *feasible* if the constraint set is nonempty, and *bounded* if the minimal value of c^Tx on the constraint set is finite.

Assuming that the problem is feasible, it follows readily that the problem is bounded if and only if $A^+Ac = c$. To see this, suppose first that the problem is bounded. If $\mathscr{N}(A) = \{0\}$ then $I-A^+A = 0$, (3.5), and so $A^+Ac = c$. Otherwise, if $0 \neq y \in \mathscr{N}(A)$, and if x_0 is in the constraint set, so is $x_0+\alpha y$ for every scalar α. Hence

$$\min_{a \leqslant Ax \leqslant b} c^Tx \leqslant c^T(x_0+\alpha y) = c^Tx_0 + \alpha c^Ty$$

which can be made arbitrarily small and negative unless c is orthogonal to every null vector of A. Thus, the problem is bounded if and only if

$$c \in \mathcal{N}^{\perp}(A) = \mathcal{R}(A^{\mathrm{T}}). \tag{2.10}$$

Since

$$A^{+}Ac \text{ is the projection of } c \text{ on } \mathcal{R}(A^{\mathrm{T}}) \tag{3.5}$$

we see that boundedness implies $A^{+}Ac = c$.

Conversely, if $A^{+}Ac = c$, then $c \in \mathcal{R}(A^{\mathrm{T}})$ so that $c = A^{\mathrm{T}}z$ for some z. Thus $c^{\mathrm{T}}x = z^{\mathrm{T}}Ax$. Each component of Ax is bounded below on the constraint set so that $z^{\mathrm{T}}Ax$ is bounded below as x ranges over the constraint set. Henceforth, we assume that $c \in \mathcal{R}(A^{\mathrm{T}})$.

We now make the additional assumption that the rows of A are linearly independent. In this case, the equation $Ax = z$ has a solution for every m-dimensional vector z, (3.12.5), and so

$$\min_{a \leqslant Ax \leqslant b} c^{\mathrm{T}}x = \min_{a \leqslant z \leqslant b} \min_{z = Ax} c^{\mathrm{T}}x.$$

The set of x's for which $z = Ax$ is of the form

$$A^{+}z + (I - A^{+}A)y$$

where y is free to vary unrestricted over n space, (3.12). Since $c \in \mathcal{R}(A^{\mathrm{T}}) = \mathcal{N}^{\perp}(A)$, we see that c is orthogonal to all null vectors of A [of which $(I - A^{+}A)y$ is one] so that

$$c^{\mathrm{T}}x = c^{\mathrm{T}}A^{+}z \qquad \text{if} \quad Ax = z.$$

Consequently,

$$\begin{aligned}\min_{a \leqslant Ax \leqslant b} c^{\mathrm{T}}x &= \min_{a \leqslant z \leqslant b} c^{\mathrm{T}}A^{+}z \\ &= \min_{a \leqslant z \leqslant b} (A^{+\mathrm{T}}c)^{\mathrm{T}}z\end{aligned}$$

and if $\tilde{z}$ minimizes the right side, any x of the form

$$\tilde{x} = A^{+}\tilde{z} + (I - A^{+}A)y$$

will minimize the left side over the constraint set.

The minimization of the right side is trivial: Denote the components of a, b, $A^{+\mathrm{T}}c$, and z by α_i, β_i, γ_i, and ζ_i, respectively. Then

$$\min_{a \leqslant z \leqslant b} (A^{+\mathrm{T}}c)^{\mathrm{T}}z = \min_{\substack{\alpha_i \leqslant \zeta_i \leqslant \beta_i \\ i = 1, \ldots, n}} \sum_{i=1}^{n} \gamma_i \zeta_i$$

and the components of the solution vector are clearly

$$\tilde{\zeta}_i = \begin{cases} \beta_i & \text{if} \quad \gamma_i < 0 \\ \alpha_i & \text{if} \quad \gamma_i > 0 \\ \text{anything between} & \\ \alpha_i \quad \text{and} \quad \beta_i \quad \text{(inclusive)} & \text{if} \quad \gamma_i = 0. \end{cases}$$

The present solution depends crucially upon the assumption that the rows of A are linearly independent. If this assumption is violated, a modified procedure (more complicated) must be resorted to in order to solve the problem (Ben-Israel and Robers [1]).

(3.13) Exercises

(3.13.1) If A and B are matrices with the same number of rows, then

$$\mathscr{R}(A) \subseteq \mathscr{R}(B) \qquad \text{if and only if} \quad BB^+A = A.$$

(3.13.2) If A and B are matrices with the same number of rows, the matrix equation

$$BX = A$$

has a solution if and only if $\mathscr{R}(A) \subseteq \mathscr{R}(B)$. In this case, any matrix of the form

$$X = B^+A + (I - B^+B)Y$$

(where Y has the same number of rows and columns as X) is a solution.

The solution is unique if and only if $B^+B = I$ [i.e., $\mathscr{N}(B) = \{0\}$].

(3.13.3) If M is nonsingular and A is rectangular with the same number of rows then

$$(MA)^+(MA) = A^+A.$$

If N is nonsingular and A is rectangular with the same number of columns,

$$(AN)(AN)^+ = AA^+.$$

(3.13.4) The matrix equation

$$AXB = C$$

has a solution in X if and only if

$$AA^+CB^+B = C.$$

In this case, the general solution is

$$X = A^+CB^+ + M - A^+AMBB^+$$

where M is arbitrary (same number of rows and columns as X).

(3.13.5) For any matrix A,

$$A^+ = WAY$$

where W and Y are, respectively, any solutions to

$$WAA^{\mathrm{T}} = A^{\mathrm{T}}$$

and

$$A^{\mathrm{T}}AY = A^{\mathrm{T}}$$

(Decell, [1]).

(3.13.6) A matrix is said to be normal if it commutes with its transpose. In this case, show that

$$A^+A = AA^+.$$

(3.13.7) If T is orthogonal, $(AT)^+ = T^{\mathrm{T}}A^+$.

(3.13.8) If $\mathscr{R}(B) \subseteq \mathscr{R}(A)$, then among those matrices X that satisfy

$$AX = B,$$

the one which minimizes the trace of $X^{\mathrm{T}}X$ is $\hat{X} = A^+B$.

If $AZ = B$, then

$$\operatorname{trace}(Z^{\mathrm{T}}Z) > \operatorname{trace}(\hat{X}^{\mathrm{T}}\hat{X}) \qquad \text{if} \quad Z \neq \hat{X}.$$

(3.13.9) The following conditions are equivalent:

(a) $XH^+ = 0$,
(b) $XH^{\mathrm{T}} = 0$,
(c) $XH^+H = 0$.

(3.13.10) If P is a projection, and if H is any rectangular matrix (having the right number of columns) then

$$(\bar{H}^{\mathrm{T}}\bar{H})^+ = P(\bar{H}^{\mathrm{T}}\bar{H})^+ = (\bar{H}^{\mathrm{T}}\bar{H})^+P, \qquad P\bar{H}^+ = \bar{H}^+,$$

and

$$\bar{H}^+ = (\bar{H}^{\mathrm{T}}\bar{H})^+H^{\mathrm{T}}$$

where

$$\bar{H} = HP.$$

(3.14) Exercises

(3.14.1) Let $h_1, h_2, \ldots h_n, \ldots$ be any collection of vectors and define

$$A_0 = I$$

$$A_n = \begin{cases} A_{n-1} & \text{if } h_n \text{ is a linear combination of } h_1, h_2, \ldots, h_{n-1} \\ A_{n-1} - \dfrac{(A_{n-1}h_n)(A_{n-1}h_n)^{\mathrm{T}}}{h_n^{\mathrm{T}} A_{n-1} h_n} & \text{otherwise.} \end{cases}$$

Show that for each n

(a) A_n is the projection on $\mathscr{L}^{\perp}(h_1, \ldots, h_n)$.
(b) $A_n h_{n+1}$ is the (unnormalized) $(n+1)$st vector in the Gramm–Schmidt orthogonalization of $h_1, h_2, \ldots, h_{n+1}$.

(3.14.2) **Stationary Probability for a Markov Chain**

A matrix P is called a stochastic matrix if its entries are nonnegative and its row sums are unity. Such matrices can be used to represent the one step transition probabilities for homogeneous Markov chains. If the process is *ergodic* (a condition which can be deduced by visual inspection of the transition matrix, Feller [1, Chap. XV]) then

$$\lim_{n\to\infty} (P^{\mathrm{T}})^n = (x \,\vdots\, x \,\vdots\, \cdots \,\vdots\, x)$$

where x is the unique *probability vector* (a vector having nonnegative components which sum to unity) satisfying

$$P^{\mathrm{T}}x = x.$$

This probability vector is the *steady state* probability distribution for the Markov chain. The ith component of x represents the (steady state) probability that the process is in state i at any given instant.

Using these facts as a starting point, show that

(a) $y = P^{\mathrm{T}}y$ only if y is a multiple of x.
(b) The row vectors of $I - P^{\mathrm{T}}$ are linearly dependent.

Denoting the columns of $I - P$ by $q_1, q_2, \ldots, q_N$, show that

(c) $q_1, q_2, \ldots, q_{N-1}$ are linearly independent.
(d) If $A_0 = I$,

$$A_n = A_{n-1} - (A_{n-1}q_n)(A_{n-1}q_n)^{\mathrm{T}} / q_n^{\mathrm{T}} A_{n-1} q_n \qquad n = 1, 2, \ldots, N-1$$

and u is the N-dimensional vector whose components are all ones, then

$$x = A_{N-1}u/u^{\mathrm{T}}A_{N-1}u.$$

[*Hint:* A_{N-1} is the projection on $\mathscr{L}^{\perp}(q_1, q_2, \ldots, q_{N-1})$; cf. Decell and Odell [1].]

In (3.6.1) and (3.6.2) we showed how a symmetric matrix and its pseudoinverse can be represented in terms of its eigenvalues and eigenvectors. Using the theory which we have developed thus far, we can deduce an analogous result for an arbitrary rectangular matrix A in terms of the eigenvalues and eigenvectors of $A^{\mathrm{T}}A$ and AA^{T} (see Good [1]).

We begin by reminding the reader that any matrix of the form $A^{\mathrm{T}}A$ has a unique symmetric square root, which has the explicit representation

$$(A^{\mathrm{T}}A)^{1/2} = TD^{1/2}T^{\mathrm{T}}$$

where T is the orthogonal matrix which reduces $A^{\mathrm{T}}A$ to the diagonal matrix D,

$$A^{\mathrm{T}}A = TDT^{\mathrm{T}}$$

and $D^{1/2}$ is obtained by taking the (positive) square root of D's (nonnegative) diagonal entries. [D's entries are clearly nonnegative since

$$D = T^{\mathrm{T}}A^{\mathrm{T}}AT$$

so that for any x

$$x^{\mathrm{T}}Dx = \|ATx\|^2 \geqslant 0.$$

In particular, if $D = \operatorname{diag}(d_1, \ldots, d_n)$ and x has all zero components except for a 1 in the jth place, $x^{\mathrm{T}}Dx = d_j \geqslant 0$.]

(3.15) Theorem: Let A be an $n \times m$ matrix and let L be the $r \times r$ diagonal matrix of (AA^{T})'s nonzero eigenvalues arranged in arbitrary order.

Then there is an $n \times r$ matrix P and an $r \times m$ matrix Q such that the following conditions hold:

(3.15.1) $A = PL^{1/2}Q.$

(3.15.2) $AA^{\mathrm{T}} = PLP^{\mathrm{T}}.$

(3.15.3) $AA^{+} = PP^{\mathrm{T}}.$

(3.15.4) $P^{\mathrm{T}}P = I.$

(3.15.5) $A^{\mathrm{T}}A = Q^{\mathrm{T}}LQ.$

(3.15.6) $A^{+}A = Q^{\mathrm{T}}Q.$

(3.15.7) $QQ^{\mathrm{T}} = I.$

Comment: By (3.15.4) and (3.15.7), the columns of P are orthonormal as are the rows of Q. By (3.15.2) and (3.15.4), $AA^{\mathrm{T}}P = PL$ which shows that the columns of P are eigenvectors of AA^{T}, the jth column being an eigenvector associated with the jth diagonal entry λ_j of L. By (3.15.5) and (3.15.7), the same goes for the jth column of Q^{T} (which is the jth row of Q). This result is often referred to as the *singular decomposition theorem.*

Proof: By (3.11.3) and (3.11.7)

$$AA^{+} = (AA^{\mathrm{T}})^{1/2}[(AA^{\mathrm{T}})^{1/2}]^{+}$$

so by (3.9.2)

(3.15.8) $$A = (AA^{\mathrm{T}})^{1/2}[(AA^{\mathrm{T}})^{1/2}]^{+}A.$$

By virtue of (3.7.1)

(3.15.9) $$AA^{\mathrm{T}} = TDT^{\mathrm{T}}$$

where

$$D = \operatorname{diag}(\lambda_1, \lambda_2, \ldots, \lambda_r, \overbrace{0, 0, \ldots, 0}^{n-r}),$$

$\lambda_j > 0$, and T is an orthogonal matrix. Let us express T and D as partitioned matrices:

$$\overset{n}{n T} = n\big[\overset{r}{P} \,\vdots\, \overset{n-r}{P_0}\big]$$

$$D = \begin{array}{c} r \\ n-r \end{array}\!\!\left[\begin{array}{c:c} \overset{r}{L} & \overset{n-r}{0} \\ \hdashline 0 & 0 \end{array}\right]$$

where

$$L = \operatorname{diag}(\lambda_1, \lambda_2, \ldots, \lambda_r).$$

Then (3.15.9) can be rewritten as

(3.15.10) $$AA^{\mathrm{T}} = PLP^{\mathrm{T}}$$

which proves (3.15.2). Define

(3.15.11) $$Q = P^{\mathrm{T}}[(AA^{\mathrm{T}})^{1/2}]^{+}A.$$

Since

$$(AA^{\mathrm{T}})^{1/2} = TD^{1/2}T^{\mathrm{T}}$$

and since

$$D^{1/2} = \left[\begin{array}{c:c} L^{1/2} & 0 \\ \hdashline 0 & 0 \end{array}\right]$$

we can write

$$(AA^{T})^{1/2} = PL^{1/2}P^{T}. \tag{3.15.12}$$

Therefore, by (3.15.11)

$$\begin{aligned} PL^{1/2}Q &= (PL^{1/2}P^{T})[(AA^{T})^{1/2}]^{+}A \\ &= A. && \text{(3.15.12) and (3.15.8)} \end{aligned}$$

This proves (3.15.1).

$$\begin{aligned} AA^{+} &= (AA^{T})(AA^{T})^{+} && \text{(3.11.7)} \\ &= (TDT^{T})(TD^{+}T^{T}) = TDD^{+}T^{T} \\ &= PP^{T} \end{aligned}$$

which proves (3.15.3). Since T is an orthogonal matrix, its columns are orthonormal, so $P^{T}P = I$ which establishes (3.15.4).

$$\begin{aligned} Q^{T}LQ &= A^{T}[(AA^{T})^{1/2}]^{+}PLP^{T}[(AA^{T})^{1/2}]^{+}A && \text{(3.15.11)} \\ &= A^{T}[(AA^{T})^{1/2}]^{+}(AA^{T})[(AA^{T})^{1/2}]^{+}A && \text{(3.15.12)} \\ &= A^{T}[(AA^{+})]A && \text{(3.11.3) and (3.11.7)} \\ &= A^{T}A \end{aligned}$$

which proves (3.15.5).

$$Q^{T}Q = A^{T}[(AA^{T})^{1/2}]^{+}PP^{T}[(AA^{T})^{1/2}]^{+}A. \qquad \text{(3.15.11)}$$

By virtue of (3.15.9)

$$\begin{aligned} (AA^{T})^{+} &= TD^{+}T^{T} \\ &= PL^{-1}P^{T} \end{aligned}$$

and so by (3.11.3)

$$[(AA^{T})^{1/2}]^{+} = [(AA^{T})^{+}]^{1/2} = PL^{-1/2}P^{T}.$$

Thus

$$\begin{aligned} Q^{T}Q &= A^{T}(PL^{-1/2}P^{T})(PP^{T})(PL^{-1/2}P^{T})A \\ &= A^{T}PL^{-1}P^{T}A \\ &= A^{T}(AA^{T})^{+}A \\ &= A^{+}A. && \text{(3.8.3)} \end{aligned}$$

This proves (3.15.6). Finally, by (3.15.11)

$$
\begin{aligned}
QQ^{\mathrm{T}} &= P^{\mathrm{T}}[(AA^{\mathrm{T}})^{1/2}]^{+}AA^{\mathrm{T}}[(AA^{\mathrm{T}})^{1/2}]^{+}P \\
&= P^{\mathrm{T}}(AA^{+})P && \text{(3.11.3) and (3.11.7)} \\
&= P^{\mathrm{T}}(PP^{\mathrm{T}})P && \text{(3.15.3)} \\
&= I. \quad \blacksquare && \text{(3.15.4)}
\end{aligned}
$$

(3.15.13) **Exercise:** The nonzero eigenvalues of $A^{\mathrm{T}}A$ and AA^{T} are identical.

(3.15.14) **Exercise:** $A^{+} = Q^{\mathrm{T}}L^{-1/2}P^{\mathrm{T}}, \quad Q^{+} = Q^{\mathrm{T}}, \quad P^{+} = P^{\mathrm{T}}.$

(3.16) Exercise: There exists a representation for A of the form (3.16.1)

(3.16.1) $$A = \sum_j \lambda_j^{1/2} p_j q_j^{\mathrm{T}}$$

where the λ_j's are the nonzero eigenvalues of $A^{\mathrm{T}}A$ (or AA^{T}) repeated according to their multiplicity,

(3.16.2) $$AA^{\mathrm{T}}p_j = \lambda_j p_j; \qquad A^{\mathrm{T}}Aq_j = \lambda_j q_j$$

and

(3.16.3) $$p_i^{\mathrm{T}}p_j = \begin{Bmatrix} 0 & \text{if} \quad i \neq j \\ 1 & \text{if} \quad i = j \end{Bmatrix} = q_i^{\mathrm{T}}q_j.$$

The representation is not necessarily unique, but not all such representations [with p_j and q_j satisfying (3.16.2)–(3.16.3)] are valid.

(3.17) Exercise: $A^{+} = \sum_j \lambda_j^{-1/2} q_j p_j^{\mathrm{T}}$ if A satisfies (3.16.1)–(3.16.3).

(3.18) Exercise: Any matrix A can be represented as a linear combination of partial isometries:

$$A = \sum \lambda_j^{1/2} U(\lambda_j)$$

where the λ_j's are the *distinct* nonzero eigenvalues of $A^{\mathrm{T}}A$ and

$$U(\lambda) = \lambda^{-1/2} A\{I - (A^{\mathrm{T}}A - \lambda I)^{+}(A^{\mathrm{T}}A - \lambda I)\}$$

satisfy

$$U(\lambda_j)\, U^{\mathrm{T}}(\lambda_k) = 0 \qquad \text{if} \quad k \neq j$$

$$U^{\mathrm{T}}(\lambda_j)\, U(\lambda_k) = 0 \qquad \text{if} \quad k \neq j$$

$$A^+ = \sum_j \lambda_j^{-1/2} U^{\mathrm{T}}(\lambda_j).$$

[*Hint:* $I-(A^{\mathrm{T}}A-\lambda I)^+(A^{\mathrm{T}}A-\lambda I)$ is the projection on the null space of $A^{\mathrm{T}}A-\lambda I$, which is spanned by the eigenvectors (if any) of $A^{\mathrm{T}}A$ associated with λ. Then use (3.16.1).] See Ben-Israel and Charnes [1], Penrose [1], and Golub and Kahan [1] for related work.

(3.19) Exercise *Iterative method for finding the dominant eigenvalue and associated eigenvector for AA^{T} and $A^{\mathrm{T}}A$.*

Let

$$y_0 = Ax_0/\|Ax_0\|$$

$$x_{n+1} = A^{\mathrm{T}}y_n/\|A^{\mathrm{T}}y_n\|$$

$$y_{n+1} = Ax_{n+1}/\|Ax_{n+1}\| \qquad n = 0, 1, \ldots.$$

Then

$$\lim_{n\to\infty} x_n = q \qquad \text{and} \qquad \lim_{n\to\infty} y_n = p \quad \text{exist}$$

provided that x_0 is not orthogonal to $\mathcal{N}(A^{\mathrm{T}}A-\lambda_1 I)$ where λ_1 is the largest eigenvalue of $A^{\mathrm{T}}A$.

Furthermore,

$$AA^{\mathrm{T}}p = \lambda_1 p \qquad \text{and} \qquad A^{\mathrm{T}}Aq = \lambda_1 q$$

so that

$$\lambda_1 = \lim_{n\to\infty} \|A^{\mathrm{T}}y_n\|^2$$

$$= \lim_{n\to\infty} \|Ax_n\|^2.$$

Chapter IV

PSEUDOINVERSES OF PARTITIONED MATRICES AND SUMS AND PRODUCTS OF MATRICES

If $c_1, c_2, \ldots, c_m$ are a collection of vectors in an n-dimensional space, we can write the $n \times m$ matrix

$$C_m = (c_1 \vdots c_2 \vdots \cdots \vdots c_m)$$

whose jth column is c_j, in terms of C_{m-1} and c_m:

(4.1) $$C_m = (C_{m-1} \vdots c_m) \qquad (m = 2, 3, \ldots).$$

The pseudoinverse of C_1 is easy to compute:

(4.2) $$C_1^{+} = c_1^{\mathrm{T}}/c_1^{\mathrm{T}}c_1 \qquad (3.11.9)$$

and so, if we can develop a convenient relationship between C_m^{+} and C_{m-1}^{+}, we will have a nice computational procedure for pseudoinverting a rectangular matrix "a column at a time":

(4.3) Theorem (Greville [2]): If

(4.3.1) $$C_{m+1} = (C_m \vdots c_{m+1})$$

then

(4.3.2)
$$C_{m+1}^{+} = \left(\frac{C_m^{+}[I - c_{m+1} k_{m+1}^{\mathrm{T}}]}{k_{m+1}^{\mathrm{T}}} \right)$$

where

(4.3.3)
$$k_{m+1} = \begin{cases} \dfrac{(I - C_m C_m^{+}) c_{m+1}}{\|(I - C_m C_m^{+}) c_{m+1}\|^2} & \text{if } (I - C_m C_m^{+}) c_{m+1} \neq 0 \\[2ex] \dfrac{C_m^{+\mathrm{T}} C_m^{+} c_{m+1}}{1 + \|C_m^{+} c_{m+1}\|^2} & \text{otherwise.} \end{cases}$$

Comment: $(I - C_m C_m^{+}) c_{m+1}$ is zero if and only if $C_m C_m^{+} c_{m+1} = c_{m+1}$ [i.e., if and only if c_{m+1} is in the range of C_m, (3.7.5)]. The range of C_m is spanned by $c_1, \ldots, c_m$ (why?) and so k_{m+1} is defined by the first part of (4.3.3) if and only if c_{m+1} is not a linear combination of $c_1, \ldots, c_m$.

Proof: The proof is a straightforward, though tedious verification that the right side of (4.3.2) satisfies the conditions of (3.9). We leave the details to the reader. (Greville's original proof was constructive. The interested reader will find it instructive, as well.) ■

(4.4) Application to Stepwise Regression

It is very common for an experimentalist to observe samples of a function of an independent variable (e.g., time) whose functional form is not known *a priori* and for him to wish to model the behavior of this function (e.g., for the purpose of prediction). Typically, the experimenter has in mind a family of functions, $\varphi_1(\tau), \varphi_2(\tau), \ldots, \varphi_m(\tau)$ and the data is modeled by choosing an appropriate set of weights and representing (approximating) the observed data as a linear combination of the φ's.

To be more explicit, if the observations denoted by $\zeta_1, \zeta_2, \ldots, \zeta_n$ are made at times $\tau_1, \tau_2, \ldots, \tau_n$ and if the points (τ_j, ζ_j) are plotted on cartesian coordinates, the problem boils down to choosing a set of scalars $\xi_1, \xi_2, \ldots, \xi_m$ so that the graph of the function

$$\sum_{j=1}^{m} \xi_j \varphi_j(\tau)$$

plotted as a function of τ, comes as close as possible to the given data points.

The most popular method for choosing the weights is (for good cause) the method of least squares, in which the ξ_j's are chosen to minimize the sums of the squares of the (vertical) distances from the data points to the curve.

In mathematical terms, the ξ_j's are chosen to minimize

$$\sum_{k=1}^{n} \left(\zeta_k - \sum_{j=1}^{m} \xi_j \varphi_j(\tau_k) \right)^2. \tag{4.4.1}$$

If a vector notation is used, so that

$$z = \begin{pmatrix} \zeta_1 \\ \vdots \\ \zeta_n \end{pmatrix} \qquad x = \begin{pmatrix} \xi_1 \\ \vdots \\ \xi_m \end{pmatrix} \qquad c_j = \begin{pmatrix} \varphi_j(\tau_1) \\ \vdots \\ \varphi_j(\tau_n) \end{pmatrix}$$

and

$$C_m = (c_1 \vdots c_2 \cdots \vdots c_m) \qquad \text{(an } n \times m \text{ matrix)}$$

then the problem of choosing $\xi_1, \ldots, \xi_m$ to minimize (4.4.1) is the same as choosing x to minimize

$$\|z - C_m x\|^2. \tag{4.4.2}$$

In (3.4) we showed that

$$\hat{x}^{(m)} = C_m^{+} z \tag{4.4.3}$$

always minimizes (4.4.2) and is in fact the vector of minimum norm which does the job. Suppose $\hat{x}^{(m)}$ is computed and that the residual sum of squares

$$\|z - C_m \hat{x}^{(m)}\|^2$$

(which measures the degree of fit between the data and the model) is unacceptably large.

The standard response to such a situation is to augment the family of functions $\varphi_1(\cdot), \varphi_2(\cdot), \ldots, \varphi_m(\cdot)$ by adding an $(m+1)$st function $\varphi_{m+1}(\cdot)$ and then choosing the $m+1$ weights $\xi_1, \xi_2, \ldots, \xi_{m+1}$ to minimize the new residual sum of squares

$$\sum_{k=1}^{n} \left(\zeta_k - \sum_{j=1}^{m+1} \xi_j \varphi_j(\tau_k) \right)^2$$

which can be expressed in vector-matrix notation as

$$\|z - C_{m+1} x\|^2$$

where z is as defined before,

$$C_{m+1} = (C_m \vdots c_{m+1}) \qquad c_{m+1} = \begin{pmatrix} \varphi_{m+1}(\tau_1) \\ \vdots \\ \varphi_{m+1}(\tau_n) \end{pmatrix}$$

and x is now an $m+1$-dimensional vector,

$$x = \begin{pmatrix} \xi_1 \\ \vdots \\ \xi_{m+1} \end{pmatrix}.$$

The minimum norm solution is

$$\hat{x}^{(m+1)} = C_{m+1}^{+} z \tag{4.4.4}$$

and the results of (4.3) show how $\hat{x}^{(m+1)}$ is related to $\hat{x}^{(m)}$:

$$\hat{x}^{(m+1)} = C_{m+1}^{+} z = \begin{pmatrix} C_m{}^{+} z - C_m{}^{+} c_{m+1} (k_{m+1}^{\mathrm{T}} z) \\ \hdashline k_{m+1}^{\mathrm{T}} z \end{pmatrix}$$

$$= \begin{pmatrix} \hat{x}^{(m)} - (k_{m+1}^{\mathrm{T}} z) C_m{}^{+} c_{m+1} \\ \hdashline k_{m+1}^{\mathrm{T}} z \end{pmatrix}.$$

(4.5) Exercise: Let $\hat{z}^{(m)} = C_m \hat{x}^{(m)}$ and $e^{(m)} = \|z - \hat{z}^{(m)}\|^2$.

(a) $e^{(m+1)} \leqslant e^{(m)}$ with strict inequality holding unless c_{m+1} is a linear combination of $c_1, \ldots, c_m$ or unless $c_{m+1} \perp (z - \hat{z}^{(m)})$.

(b) $$e^{(m+1)} = (1 - \rho_{z,m+1|m}^2) e^{(m)},$$

$$\rho_{z,m+1|m} = [(z - \hat{z}^{(m)})^{\mathrm{T}} \tilde{c}_{m+1}][\|z - \hat{z}^{(m)}\| \cdot \|\tilde{c}_{m+1}\|]^{+}$$

and

$$\tilde{c}_{m+1} = (I - C_m C_m{}^{+}) c_{m+1}.$$

$\rho_{z,m+1|m}$ is called the *partial* correlation coefficient between z and c_{m+1}, given $c_1, \ldots, c_m$. Interpret the partial correlation coefficient geometrically.

(c) $e^{(m)} = [\prod_{j=1}^{m} (1 - \rho_{z,j|j-1}^2)] \|z\|^2$.

(d) The *multiple* correlation coefficient of z on $c_1, \ldots, c_m$ is defined to be

$$r_{z|m} = z^{\mathrm{T}} \hat{z}^{(m)} / (\|z\| \cdot \|\hat{z}^{(m)}\|).$$

Interpret the multiple correlation coefficient geometrically and show that

$$\prod_{j=1}^{m} (1 - \rho_{z,j|j-1}^2) = 1 - r_{z|m}^2.$$

A word of caution is in order with regard to the interpretation of (4.5a). The object of modeling data is generally predictive in nature. Therefore, the fact that the residual sum of squares can be reduced by the addition of another regressor is not sufficient cause to add regressors *ad infinitum*. A fine line must be tread between parsimonious modeling (using few regressors) and getting a good fit. In many statistical contexts, underspecifying the number of regressors results in biased estimates. Over specification (too many regressors) results in loss of accuracy. The question, "How many regressors is enough?" is partially answered by the theory of the analysis of variance and covariance, under certain assumptions. The question, "How should I choose my family of regressors to begin with?" is much more difficult and not easily answered in quantitative terms.

The results of (4.3) lead directly to an important representation for pseudoinverses of certain types of matrix sums:

(4.6) Theorem: Let $c_1, c_2, \ldots$ be a collection of n-dimensional vectors and let

$$S_m = \sum_{j=1}^{m} c_j c_j^{\mathrm{T}} = C_m C_m^{\mathrm{T}} \qquad (m = 1, 2, \ldots)$$

where C_m is as defined in (4.1). Then

$$\text{(4.6.1)} \quad S_{m+1}^{+} = \begin{cases} S_m^{+} + \left[\dfrac{1 + c_{m+1}^{\mathrm{T}} S_m^{+} c_{m+1}}{(c_{m+1}^{\mathrm{T}} A_m c_{m+1})^2}\right](A_m c_{m+1})(A_m c_{m+1})^{\mathrm{T}} \\ \quad - \dfrac{(S_m^{+} c_{m+1})(A_m c_{m+1})^{\mathrm{T}} + (A_m c_{m+1})(S_m^{+} c_{m+1})^{\mathrm{T}}}{c_{m+1}^{\mathrm{T}} A_m c_{m+1}} \\ \qquad \text{if } c_{m+1} \text{ is not a linear combination of } c_1, \ldots, c_m \\ S_m^{+} - \dfrac{(S_m^{+} c_{m+1})(S_m^{+} c_{m+1})^{\mathrm{T}}}{1 + c_{m+1}^{\mathrm{T}} S_m^{+} c_{m+1}} \qquad \text{otherwise} \end{cases}$$

where

$$\text{(4.6.2)} \qquad A_m = I - S_m S_m^{+}.$$

Comment: A_m is the projection on $\mathscr{N}(S_m^{\mathrm{T}}) = \mathscr{R}^{\perp}(S_m)$, (3.5). Since $S_m = C_m C_m^{\mathrm{T}}$ and

$$\mathscr{R}(C_m C_m^{\mathrm{T}}) = \mathscr{R}(C_m) \qquad (2.12)$$
$$= \mathscr{L}(c_1, c_2, \ldots, c_m) \qquad (4.3.4)$$

we see that A_m is the projection on $\mathscr{L}^{\perp}(c_1, \ldots, c_m)$, and so $c_{m+1} \in \mathscr{L}(c_1, \ldots, c_m)$ if and only if $A_m c_{m+1} = 0$. Thus, c_{m+1} is *not* a linear combination of $c_1, \ldots, c_m$ if and only if $A_m c_{m+1} \neq 0$.

Furthermore, by virtue of (3.14.1), A_m satisfies a recursion of its own [replace h_m by c_m in (3.14.1)].

Proof of (4.6)

$$S_{m+1}^{+} = (C_{m+1} C_{m+1}^{\mathrm{T}})^{+}$$

$$= (C_{m+1}^{+})^{\mathrm{T}} (C_{m+1}^{+}). \qquad (3.11.2) \text{ and } (3.8.2) \tag{4.6.3}$$

(4.6.1) results when (4.3.2) and (4.3.3) are combined in (4.6.3) and note is taken of the fact that

$$I - C_m C_m^{+} = I - S_m S_m^{+} = A_m. \quad \blacksquare$$

(4.6.4) **Exercise** [Extension of (3.14.1)]: Let

$$C_m = (c_1 \vdots c_2 \vdots \cdots \vdots c_m)$$

and

$$D_m = D_0 - D_0 C_m (C_m^{\mathrm{T}} D_0 C_m)^{+} C_m^{\mathrm{T}} D_0$$

where D_0 is an arbitrary matrix of the form $D_0 = R^{\mathrm{T}} R$. Then

(a) $D_m = R^{\mathrm{T}} Q_m R$

where Q_m is the projection on $\mathscr{L}^{\perp}(Rc_1, \ldots, Rc_m) = \mathscr{R}^{\perp}(RC_m)$.

(b)
$$D_{m+1} = \begin{cases} D_m - \dfrac{(D_m c_{m+1})(D_m c_{m+1})^{\mathrm{T}}}{c_{m+1}^{\mathrm{T}} D_m c_{m+1}} & \text{if } Rc_{m+1} \text{ is not a l.c. (linear combination) of } Rc_1, \ldots, Rc_m; \\ D_m & \text{otherwise.} \end{cases}$$

(c) If $R^{\mathrm{T}} R$ is nonsingular, Rc_{m+1} is a linear combination of $Rc_1, \ldots, Rc_m$ if and only if $D_m c_{m+1} = 0$, so that D_{m+1} is defined by the first half of the recursion if $D_m c_{m+1} \neq 0$, otherwise by the second half.

Comment: If $A_m = I - C_m C_m^{+}$ then $A_0 = I$ and

$$A_{m+1} = \begin{cases} A_m - \dfrac{(A_m c_{m+1})(A_m c_{m+1})^{\mathrm{T}}}{c_{m+1}^{\mathrm{T}} A_m c_{m+1}} & \text{if } A_m c_{m+1} \neq 0 \\ A_m & \text{otherwise.} \end{cases} \qquad (3.14.1)$$

If $R^{\mathrm{T}} R$ is nonsingular, D_m and A_m satisfy the same recursion. A_m and D_m differ because of the different initial conditions.

(4.6.5) **Exercise:** If A is symmetric, nonsingular and $h^{\mathrm{T}} A^{-1} h \neq -1$, then

$$(A + hh^{\mathrm{T}})^{-1} = A^{-1} - \frac{(A^{-1}h)(A^{-1}h)^{\mathrm{T}}}{1 + h^{\mathrm{T}} A^{-1} h}.$$

The results of (4.3) and (4.6) can be extended to the case of higher-order partitioning. The proofs are computational in nature, but are basically the same as (4.3) and (4.6). We state the results without proofs and refer the interested reader to the papers by Cline [2, 3] for details.

(4.7) Theorem

(4.7.1)
$$(U \vdots V)^{+} = \begin{pmatrix} U^{+}-U^{+}VJ \\ \hdashline J \end{pmatrix}$$

where

(4.7.2) $J = C^{+}+(I-C^{+}C)KV^{\mathrm{T}}U^{+\mathrm{T}}U^{+}(I-VC^{+})$

(4.7.3) $C = (I-UU^{+})V$

and

(4.7.4) $K = \{I+[U^{+}V(I-C^{+}C)]^{\mathrm{T}}[U^{+}V(I-C^{+}C)]\}^{-1}.$

Comment: Any matrix of the form $I+D^{\mathrm{T}}D$ is nonsingular, (2.13), so K always exists. The dimension of K is the same as the dimension of $C^{\mathrm{T}}C$ which is the same as the dimension of $V^{\mathrm{T}}V$. If U^{+} is known, $(U \vdots V)^{+}$ can be computed at the expense of inverting a square matrix the size of $V^{\mathrm{T}}V$, and finding $I-C^{+}C$, the projection on $\mathcal{N}(C)$.

The extension of (4.6) to higher-order partitions is

(4.8) Theorem

(4.8.1)
$$\begin{aligned}(UU^{\mathrm{T}}+VV^{\mathrm{T}})^{+} &= (CC^{\mathrm{T}})^{+}+[I-(VC^{+})^{\mathrm{T}}] \\ &\quad \times [(UU^{\mathrm{T}})^{+}-(UU^{\mathrm{T}})^{+}V(I-C^{+}C)KV^{\mathrm{T}}(UU^{\mathrm{T}})^{+}] \\ &\quad \times [I-VC^{+}]\end{aligned}$$

where C and K are as defined in (4.7).

Comment: We can also write

(4.8.2) $C = [I-(UU^{\mathrm{T}})(UU^{\mathrm{T}})^{+}]V$

and

(4.8.3) $K = \{I+[(I-C^{+}C)V^{\mathrm{T}}(UU^{\mathrm{T}})^{+}V(I-C^{+}C)]\}^{-1}.$

Here, U and V can be interchanged throughout the right side of (4.8.1) without altering the validity of the statement, owing to the symmetry of the left side.

(4.8.4) **Exercise:** $(U \vdots 0)^{+} = \begin{pmatrix} U^{+} \\ \hdashline 0 \end{pmatrix}.$

(4.8.5) **Exercise:** In the special case where $U = C_m$ and $V = c_{m+1}$, show that (4.7) and (4.8) yield formulas which coincide with (4.3) and (4.6).

Theorem (4.8) is extremely important if for no other reason than that it allows the development of an explicit perturbation theory for the pseudoinverse of matrices of the form $A^{\mathrm{T}}A$:

(4.9) Theorem

$$(4.9.1)\quad [H^{\mathrm{T}}H+\lambda^{-2}G^{\mathrm{T}}G]^{+} = (\bar{H}^{\mathrm{T}}\bar{H})^{+} + \lambda^{2}(I-\bar{H}^{+}H)(G^{\mathrm{T}}G)^{+}(I-\bar{H}^{+}H)^{\mathrm{T}}$$
$$- \lambda^{4}(I-\bar{H}^{+}H)[H(G^{\mathrm{T}}G)^{+}]^{\mathrm{T}}$$
$$\times QM(\lambda)Q[H(G^{\mathrm{T}}G)^{+}](I-\bar{H}^{+}H)^{\mathrm{T}}$$

where

$$(4.9.2)\qquad \bar{H} = H(I-G^{+}G) = H[I-(G^{\mathrm{T}}G)^{+}(G^{\mathrm{T}}G)]$$

$$(4.9.3)\qquad Q = I - \bar{H}\bar{H}^{+}$$

and

$$(4.9.4)\qquad M(\lambda) = [I+\lambda^{2}QH(G^{\mathrm{T}}G)^{+}H^{\mathrm{T}}Q]^{-1}.$$

Proof: Let $U = G^{\mathrm{T}}/\lambda$, $V = H^{\mathrm{T}}$ and apply (4.8):

$$[H^{\mathrm{T}}H+\lambda^{-2}G^{\mathrm{T}}G]^{+} = (CC^{\mathrm{T}})^{+} + \lambda^{2}[I-(H^{\mathrm{T}}C^{+})]^{\mathrm{T}}$$
$$\times [(G^{\mathrm{T}}G)^{+} - \lambda^{2}(G^{\mathrm{T}}G)^{+}H^{\mathrm{T}}(I-C^{+}C)KH(G^{\mathrm{T}}G)^{+}]$$
$$\times (I-H^{\mathrm{T}}C^{+})$$

where

$$C = (I-G^{+}G)H^{\mathrm{T}} = \bar{H}^{\mathrm{T}}, \qquad I - C^{+}C = Q$$

and

$$K = [I+\lambda^{2}QH(G^{\mathrm{T}}G)^{+}H^{\mathrm{T}}Q]^{-1} = M(\lambda).$$

But,

$$(I-H^{\mathrm{T}}C^{+})^{\mathrm{T}} = (I-H^{\mathrm{T}}\bar{H}^{\mathrm{T}+})^{\mathrm{T}} = I - \bar{H}^{+}H$$

and

$$G^{\mathrm{T}+}H^{\mathrm{T}} = (HG^{+})^{\mathrm{T}}.$$

Finally, if $I+A$ is nonsingular, then

$$(I+A)^{-1} = I - (I+A)^{-1}A$$

so that

$$(I-C^{+}C)K \equiv QM(\lambda) = QM(\lambda)Q$$

and (4.9.1) follows directly. ■

A scalar function $\varphi(\cdot)$, of a real variable λ, is said to be $O(\lambda^n)$ as $\lambda \to 0$ if $\varphi(\lambda)/\lambda^n$ is bounded as $\lambda \to 0$. A matrix valued function is $O(\lambda^n)$ if each entry of the matrix is $O(\lambda^n)$.

(4.9.5) **Corollary**

(a) $$M(\lambda) = I + O(\lambda^2) \quad \text{as} \quad \lambda \to 0.$$

(b) $$[H^T H + (\lambda^2)^{-1} G^T G]^+ = (\bar{H}^T \bar{H})^+ + \lambda^2 (I - \bar{H}^+ H)(G^T G)^+ \times (I - \bar{H}^+ H)^T + O(\lambda^4) \qquad \text{as} \quad \lambda \to 0.$$

Proof: (a) $M(\lambda)$ is obviously $O(1)$, since $\lim_{\lambda \to 0} M(\lambda) = I$. Since $(I+A)^{-1} = I - A(I+A)^{-1}$, we see that $M(\lambda) - I = -\lambda^2 QH(G^T G)^+ H^T Q M(\lambda)$ so that $[M(\lambda) - I]/\lambda^2$ equals a constant matrix times $M(\lambda)$, which is bounded. Thus, $M(\lambda) - I = O(\lambda^2)$, as asserted.

(b) Follows directly. ■

(4.9.6) **Exercise:** Show that $\lim_{\lambda \to \infty} M(\lambda)$ exists, and calculate the limit. [*Hint:* Let $\varepsilon = 1/\lambda$ and apply (4.9.5b) to $[I + \varepsilon^{-2} A^T A]^{-1}$, where $A = G^{T+} H^T Q$.]

(4.9.7) **Exercise:** Under what conditions will $\lim_{\varepsilon \to 0} (H^T H + \varepsilon^2 G^T G)^+$ exist?

Comment: Exercise (4.9.7) again illustrates the extreme discontinuous nature of pseudoinverses. (See the discussion preceding (3.7.3). Also see Stewart [1] and Ben-Israel [1]).

(4.9.8) **Exercise:** (a) If $U^T U$ is nonsingular and $A(\lambda) = O(1)$ as $\lambda \to 0$, then $U^T U + \lambda A(\lambda)$ is nonsingular for all suitably small λ, and

(b) $$[U^T U + \lambda A(\lambda)]^{-1} = (U^T U)^{-1} + O(\lambda) \quad \text{as} \quad y \to 0.$$

(4.9.9) **Exercise:** Let C be an arbitrary symmetric matrix. Then

(a) $(C + \lambda^2 I)^{-1}$ and $(I + \lambda^2 C^+)^{-1}$ exist whenever λ^2 is suitably large or suitably small and so

(b) $[I + \lambda^{-2} C]^{-1} = (I - CC^+) + \lambda^2 C^+ (I + \lambda^2 C^+)^{-1}$ if λ^2 is suitably large or suitably small, and

(c) $(I + \lambda^2 C)^{-1} = (I - CC^+) + C^+ (C^+ + \lambda^2 I)^{-1}$ if λ^2 is suitably large or suitably small.

(d) $[I + \lambda^{-2} C]^{-1} = (I - CC^+) - \sum_{j=1}^{n} (-\lambda^2)^j (C^+)^j + O(\lambda^{2n+2})$ as $\lambda \to 0$.

[*Hint:* Write $C = T^T D T$ where T is orthogonal and D is diagonal, and use (3.6).]

(4.10) The Concept of Rank

If $\mathscr{L}$ is a linear manifold in a Euclidean n-space, the *dimension of* $\mathscr{L}$ [abbreviated $\dim(\mathscr{L})$] is defined to be the maximum number of vectors in $\mathscr{L}$ that can be chosen linearly independent of one another. A fundamental fact that we shall take for granted is that *any* basis for $\mathscr{L}$ has exactly r linearly independent vectors, where $r = \dim(\mathscr{L})$.

(4.10.1) **Exercise:** If $\mathscr{L}_1 \subseteq \mathscr{L}_2$, then

$$\dim(\mathscr{L}_2 - \mathscr{L}_1) = \dim(\mathscr{L}_2) - \dim(\mathscr{L}_1).$$

If A is any matrix, the *rank of* A [abbreviated $\mathrm{rk}(A)$] is defined to be the dimension of A's range:

$$\mathrm{rk}(A) = \dim[\mathscr{R}(A)].$$

Several properties of rank play an important role in the theory of pseudoinverses for products and in the statistical applications of Chapter VI. They are straightforward consequences of already established results and are left as exercises:

(4.10.2) **Exercise**

(a) For any matrix A

$$\mathrm{rk}(A) = \mathrm{rk}(A^{\mathrm{T}}A) = \mathrm{rk}(A^{\mathrm{T}}) = \mathrm{rk}(AA^{\mathrm{T}}).$$

(b) For any matrices A and B (of the right size)

$$\mathrm{rk}(AB) = \mathrm{rk}(A^{+}AB) = \mathrm{rk}(ABB^{+}).$$

(c) $\mathrm{rk}(AB) \leqslant \min[\mathrm{rk}(A), \mathrm{rk}(B)]$.

(d) If $\mathrm{rk}(A) = \mathrm{rk}(AB)$ then $\mathscr{R}(A) = \mathscr{R}(AB)$ and the equation

$$ABX = A$$

has a solution in X.

(4.10.3) **Exercise:** (a) If P is a projection, $\mathrm{rk}(P) = \mathrm{trace}(P)$.
(b) If P_0, P_1, and P_2 are projections with $\mathscr{R}(P_1) \subseteq \mathscr{R}(P_0)$, $\mathscr{R}(P_2) \subseteq \mathscr{R}(P_0)$ and $P_1 P_2 = 0$ then

$$P_0 = P_1 + P_2$$

if and only if $\mathrm{rk}(P_0) = \mathrm{rk}(P_1) + rk(P_2)$ (c.f. Cochran's theorem, Scheffé [1]).

We now turn our attention to the question of pseudoinverses for products of matrixes. If A and B are nonsingular, $(AB)^{-1} = B^{-1}A^{-1}$, but it is not generally true that $(AB)^+ = B^+A^+$ as evidenced by the example

$$A = (1 \quad 0) \qquad B = \begin{pmatrix} 1 \\ 1 \end{pmatrix} \qquad (AB)^+ = 1 \qquad B^+A^+ = \frac{1}{2}.$$

Greville [3] has found necessary and sufficient conditions for $(AB)^+ = B^+A^+$. These we state as the following.

(4.11) Theorem: $(AB)^+ = B^+A^+$ if and only if

(4.11.1) $$\mathscr{R}(BB^TA^T) \subseteq \mathscr{R}(A^T)$$

and

(4.11.2) $$\mathscr{R}(A^TAB) \subseteq \mathscr{R}(B).$$

Proof: Since $A^+A = A^TA^{T+}$, (4.11.1) holds if and only if the equation

(4.11.3) $$A^+ABB^TA^T = BB^TA^T. \tag{3.13.1}$$

By the same token, (4.11.2) holds if and only if

(4.11.4) $$BB^+A^TAB = A^TAB.$$

We will show that (4.11.3) and (4.11.4) are necessary and sufficient for $(AB)^+ = B^+A^+$:

Suppose (4.11.3) and (4.11.4) hold. Multiply (4.11.3) on the left by B^+ and on the right by $(AB)^{T+}$:

$$\begin{aligned} &B^+[A^+A(BB^TA^T)](B^TA^T)^+ \\ &\quad = B^+A^+(AB)[(AB)^+(AB)]^T \\ &\quad = B^+A^+(AB) \end{aligned}$$

while

$$\begin{aligned} &B^+[BB^TA^T](AB)^{T+} \\ &\quad = B^+BB^TA^T(AB)^{T+} \\ &\quad = (AB)^T(AB)^{T+} \\ &\quad = (AB)^+(AB) \end{aligned}$$

so that if (4.11.3) holds

(4.11.5) $$B^+A^+(AB) = (AB)^+(AB).$$

By the same token, if both sides of (4.11.4) are premultiplied by $(AB)^{T+}$ and postmultiplied by A^+, we find that

(4.11.6) $$(AB)B^+A^+ = (AB)(AB)^+.$$

The right-hand sides of (4.11.5) and (4.11.6) are symmetric and so B^+A^+ satisfies (3.9.1). We are done if we can show that B^+A^+ satisfies (3.9.2) and (3.9.3) as well:

If (4.11.5) is premultiplied on both sides by AB, we see that B^+A^+ does indeed satisfy (3.9.2). (3.9.3) is a bit more subtle:

Since

$$\begin{aligned} B^+A^+ &= (B^+BB^+)(A^+AA^+) \\ &= (B^+B^{+\mathrm{T}})(B^{\mathrm{T}}A^{\mathrm{T}})(A^{+\mathrm{T}}A^+) \end{aligned}$$

(4.10.2) tells us that

$$\text{(4.11.7)} \qquad \operatorname{rk}(B^+A^+) \leqslant \operatorname{rk}(B^{\mathrm{T}}A^{\mathrm{T}}) = \operatorname{rk}(AB).$$

On the other hand, (4.10.2b) asserts that

$$\text{(4.11.8)} \qquad \begin{aligned} \operatorname{rk}(AB) &= \operatorname{rk}[(AB)^+(AB)] \\ &= \operatorname{rk}[(B^+A^+)(AB)] && \text{(4.11.5)} \\ &\leqslant \operatorname{rk}(B^+A^+). && \text{(4.10.2c)} \end{aligned}$$

If (4.11.7) and (4.11.8) are combined, we find that

$$\text{(4.11.9)} \qquad \operatorname{rk}(B^+A^+) = \operatorname{rk}(B^+A^+AB)$$

so that the equation

$$\text{(4.11.10)} \qquad B^+A^+ABX = B^+A^+$$

has a solution in X. (4.10.2d)

Premultiply (4.11.10) by AB and apply (4.11.6) to deduce

$$\text{(4.11.11)} \qquad \begin{aligned} &(AB)(B^+A^+)(AB)X = \\ &(AB)(AB)^+(AB)X = \\ &\qquad (AB)X = (AB)(B^+A^+). \end{aligned}$$

Substituting the last expression for ABX in (4.11.10), we see that

$$(B^+A^+)(AB)(B^+A^+) = B^+A^+$$

which establishes (3.9.3) and shows that $B^+A^+ = (AB)^+$ if (4.11.3) and (4.11.4) hold.

To prove the converse, suppose that $(AB)^+ = B^+A^+$: Then

$$\begin{aligned} (AB)^{\mathrm{T}} &= [(AB)(AB)^+(AB)]^{\mathrm{T}} \\ &= (AB)^+(AB)(AB)^{\mathrm{T}} \\ &= B^+A^+(AB)(AB)^{\mathrm{T}}. \end{aligned}$$

If the left and right sides are premultiplied by $ABB^{\mathrm{T}}B$ and use is made of the identity

$$\begin{aligned} B^{\mathrm{T}} &= (BB^{+}B)^{\mathrm{T}} \\ &= B^{\mathrm{T}}BB^{+} \end{aligned}$$

we find that

$$ABB^{\mathrm{T}}B(AB)^{\mathrm{T}} = ABB^{\mathrm{T}}BB^{+}A^{+}(AB)(AB)^{\mathrm{T}}$$

or equivalently

(4.11.12) $$ABB^{\mathrm{T}}(I-A^{+}A)\,BB^{\mathrm{T}}A^{\mathrm{T}} = 0.$$

Since $H^{\mathrm{T}}H = 0$ implies $H = 0$, (4.11.2) tells us that

$$(I-A^{+}A)\,BB^{\mathrm{T}}A^{\mathrm{T}} = 0 \qquad [\text{take } H = (I-A^{+}A)\,BB^{\mathrm{T}}A^{\mathrm{T}}]$$

which is the same as (4.11.3). Equation (4.11.4) is proved the same way, interchanging A^{T} and B throughout in the preceding proof. ■

A general representation for $(AB)^{+}$ was derived by Cline [1].

(4.12) Theorem: $(AB)^{+} = B_1{}^{+}A_1{}^{+}$

where

(4.12.1) $$B_1 = A^{+}AB$$

and

(4.12.2) $$A_1 = AB_1B_1{}^{+}.$$

Proof: Clearly

(4.12.3) $$AB = A_1B_1.$$

Furthermore,

(4.12.4) $$B_1B_1{}^{+} = A^{+}A_1$$

since

$$A^{+}A_1 = A^{+}(AB_1B_1{}^{+}) = (A^{+}A)(A^{+}A)\,BB_1{}^{+} = A^{+}ABB_1{}^{+}.$$

Similarly

(4.12.5) $$A_1{}^{+}A_1 = A^{+}A_1$$

because

$$(A_1{}^{+}A_1)(B_1B_1{}^{+}) = A_1{}^{+}(AB_1B_1{}^{+})\,B_1B_1{}^{+} = A_1{}^{+}(AB_1B_1{}^{+}) = A_1{}^{+}A_1$$

from which follows (4.12.5) after transposes are taken and note is made of (4.12.4).

It is now easy to show that A_1 and B_1 satisfy (4.11.3) and (4.11.4):

$$\begin{aligned} A_1{}^+ A_1 B_1 (A_1 B_1)^{\mathrm{T}} &= (A^+ A_1) B_1 (A_1 B_1)^{\mathrm{T}} && (4.12.5) \\ &= A^+(AB)(A_1 B_1)^{\mathrm{T}} && (4.12.3) \\ &= B_1 (A_1 B_1)^{\mathrm{T}} && (4.12.1) \end{aligned}$$

so (4.11.3) holds.

Similarly,

$$\begin{aligned} B_1 B_1{}^+ A_1{}^{\mathrm{T}}(A_1 B_1) &= (A^+ A_1) A_1{}^{\mathrm{T}}(A_1 B_1) && (4.12.4) \\ &= (A_1{}^+ A_1)(A_1{}^{\mathrm{T}})(A_1 B_1) && (4.12.5) \\ &= A_1{}^{\mathrm{T}} A_1 B_1 \end{aligned}$$

which proves (4.11.4), so that

$$(A_1 B_1)^+ = B_1{}^+ A_1{}^+.$$

The desired conclusion follows from (4.12.3). ■

(4.13) Exercise: If A is $n \times r$ and B is $r \times m$ then

$$(AB)^+ = B^+ A^+$$

if $\mathrm{rk}(A) = \mathrm{rk}(B) = r$.

(4.14) Exercise: $(AB)^+ = B^+ A^+$ if

(a) $A^{\mathrm{T}} A = I$ or

(b) $BB^{\mathrm{T}} = I$ or

(c) $B = A^{\mathrm{T}}$ or

(d) $B = A^+$.

(4.15) Exercise: If H is rectangular and S is symmetric and nonsingular,

$$(SH)^+ = H^+ S^{-1}[I-(QS^{-1})^+(QS^{-1})]$$

where

$$Q = (I - HH^+).$$

Chapter V

COMPUTATIONAL METHODS

In recent years, a sizable literature relating to the computation of pseudo-inverses has accompanied the rebirth of interest in the theory. In this chapter, we will describe four distinct approaches to the problem. The first method is based upon the Gramm–Schmidt orthogonalization (hereafter abbreviated GSO), the second is a modification of the "old faithful," Gauss–Jordan elimination, the third is based upon the ideas of gradient projection and the last is an exotic procedure derived from the Cayley–Hamilton theorem.

(5.1) Method I (GSO, Rust, Burrus, and Schneeburger [1])

Let A be an $n \times m$ matrix of rank $k \leqslant \min(n, m)$. It is always possible to rearrange the columns of A so that the first k columns are linearly independent while the remaining columns are linear combinations of the first k.

This is the same as saying that for some *permutation* matrix, P (a square matrix of zeros and ones with exactly one nonzero entry in each row and column)

(5.1.1) $$AP = (R \mid S)$$

where R is $n \times k$ and has rank k and the columns of S are linear combinations of the columns of R:

(5.1.2) $$S = RU \quad \text{for some} \quad U. \tag{3.13.2}$$

P is an orthogonal matrix so that

$$A = (R \mid RU)P^{\mathrm{T}},$$

and

$$A^+ = P[R(I \mid U)]^+. \tag{4.14}$$

The rank of $(I \mid U)$ is the same as the rank of $(I \mid U)(I \mid U)^{\mathrm{T}} = I + UU^{\mathrm{T}}$, (4.10.2a) which is k. Therefore the rows of $(I \mid U)$ are linearly independent so that

$$[R(I \mid U)]^+ = (I \mid U)^+ R^+ \tag{4.13}$$

$$= (I \mid U)^{\mathrm{T}}(I + UU^{\mathrm{T}})^{-1} R^+, \tag{3.5.2}$$

hence

(5.1.3) $$A^+ = P(I \mid U)^{\mathrm{T}}(I + UU^{\mathrm{T}})^{-1} R^+.$$

The last equation is the starting point for the computational procedure based on GSO; GSO is used to evaluate P, R^+, U, and $(I + UU^{\mathrm{T}})^{-1}$:

(a) *Evaluation of P*

Perform a GSO on the columns of A, but *do not* normalize: i.e., denote the columns of A by $a_1, a_2, \ldots, a_m$, let

$$c_1{}^* = a_1$$

$$c_j{}^* = a_j - \sum_{i \in S_j} \frac{a_j{}^{\mathrm{T}} c_i{}^*}{\|c_i{}^*\|^2} c_i{}^*$$

where

$$S_j = \{i \colon i \leqslant j-1 \text{ and } c_i{}^* \neq 0\}.$$

The vectors $c_j{}^*$ are mutually orthogonal and

$$\mathscr{L}(c_1{}^*, c_2{}^*, \ldots, c_i{}^*) = \mathscr{L}(a_1, a_2, \ldots, a_i) \qquad \text{for each} \quad i. \tag{2.8.1}$$

If the vectors $c_1{}^*, c_2{}^*, \ldots, c_m{}^*$ are permuted so that the nonzero vectors (of which there will be k) come first, the same permutation matrix applied to the vectors $a_1, a_2, \ldots, a_m$ will rearrange them so that the first k are linearly independent, while the last $m-k$ are linear combinations of the first k, since $c_j{}^* = 0$ if and only if a_j is a linear combination of the preceding a's. So, if P is any matrix for which

(5.1.4) $$(c_1{}^* \mid c_2{}^* \mid \cdots \mid c_m{}^*) P = (c_1 \mid c_2 \mid \cdots \mid c_m)$$

where

$$\|c_j\| \begin{cases} > 0 & j = 1,2,\ldots,k \\ = 0 & j = k+1,\ldots,m \end{cases}$$

then

(5.1.5) $$AP = (a_1 \,|\, a_2 \,|\, \cdots \,|\, a_m)\, P = (R \,|\, S)$$

where R is $n \times k$ of rank k and the columns of S are linear combinations of the columns of R.

(b) *Computation of R^+*

The (nonzero) vectors, $c_1, c_2, \ldots, c_k$ defined above, represent a GSO of the columns of R. If we let

(5.1.6) $$Q = \left(\frac{c_1}{\|c_1\|} \,\middle|\, \frac{c_2}{\|c_2\|} \,\middle|\, \cdots \,\middle|\, \frac{c_k}{\|c_k\|} \right)$$

then

$$\mathcal{R}(Q) = \mathcal{R}(R)$$

so by (3.13.2) there is a $k \times k$ matrix B such that

(5.1.7) $$RB = Q.$$

Indeed since R has rank k, $B = (R^{\mathrm{T}} R)^{-1} R^{\mathrm{T}} Q$. We will derive an algorithm for B in (5.1.16). Since the columns of Q are orthonormal, $Q^{\mathrm{T}} Q = I$ so that B is nonsingular ($Q^{\mathrm{T}} RB = I$), and hence

(5.1.8) $$R = QB^{-1}.$$

Exercise (4.14) applies again and we find that

(5.1.9) $$R^+ = BQ^+ = B(Q^{\mathrm{T}} Q)^+ Q^{\mathrm{T}} = BQ^{\mathrm{T}}.$$

It remains to evaluate B, U, and $(I + UU^{\mathrm{T}})^{-1}$.

(c) *Computation of B and U*

Denote the columns of R by $r_1, r_2, \ldots, r_k$ and the columns of S by $s_1, s_2, \ldots, s_{m-k}$. The vectors $(c_1, c_2, \ldots, c_k, c_{k+1}, \ldots, c_m)$ defined in (5.1.4) represent a nonnormalized GSO of $(r_1, r_2, \ldots, r_k, s_1, \ldots, s_{m-k})$. Indeed

$$c_1 = r_1$$

(5.1.10) $$c_j = r_j - \sum_{i=1}^{j-1} \frac{r_j^{\mathrm{T}} c_i}{\|c_i\|^2} c_i \qquad j = 2,\ldots,k$$

and

$$0 = c_{k+j} = s_j - \sum_{i=1}^{k} \frac{s_j^{\mathrm{T}} c_i}{\|c_i\|^2} c_i \qquad j = 1, \ldots, m-k. \tag{5.1.11}$$

From (5.1.10) it is easy to deduce (by induction on j) that

$$c_j = \sum_{i=1}^{j} \gamma_{ij} r_i \qquad j = 1, 2, \ldots, k \tag{5.1.12}$$

where

$$\gamma_{ij} = \begin{cases} 0 & i > j \\ 1 & i = j \\ -\displaystyle\sum_{\alpha=i}^{j-1} \frac{(r_j^{\mathrm{T}} c_\alpha)}{\|c_\alpha\|^2} \gamma_{i\alpha} & i < j. \end{cases} \tag{5.1.13}$$

On the other hand, (5.1.11) shows that

$$s_j = \sum_{i=1}^{k} \omega_{ij} r_i \tag{5.1.14}$$

where ω_{ij} is obtained by substituting (5.1.12) into (5.1.11):

$$\begin{aligned} s_j &= \sum_{\alpha=1}^{k} \frac{s_j^{\mathrm{T}} c_\alpha}{\|c_\alpha\|^2} \left(\sum_{i=1}^{\alpha} \gamma_{i\alpha} r_i \right) \\ &= \sum_{i=1}^{k} \left(\sum_{\alpha=i}^{k} \frac{s_j^{\mathrm{T}} c_\alpha}{\|c_\alpha\|^2} \gamma_{i\alpha} \right) r_i . \end{aligned}$$

$$\omega_{ij} = \sum_{\alpha=i}^{k} \frac{(s_j^{\mathrm{T}} c_\alpha)}{\|c_\alpha\|^2} \gamma_{i\alpha} \qquad i = 1, 2, \ldots, k; \quad j = 1, \ldots, m-k. \tag{5.1.15}$$

From (5.1.12) and (5.1.6), we see that

$$Q = \left(\frac{c_1}{\|c_1\|} \,\vdots\, \cdots \,\vdots\, \frac{c_k}{\|c_k\|} \right) = RB$$

where B is the $k \times k$ matrix whose (i,j)th entry is

$$\beta_{ij} = \gamma_{ij} / \|c_j\| \tag{5.1.16}$$

while from (5.1.14) we conclude that

$$S = RU \tag{5.1.17}$$

where U is the $k \times m-k$ matrix whose (i,j)th entry is ω_{ij}.

Notice that (5.1.13) defines γ_{ij} in terms of $\gamma_{ij-1}, \gamma_{ij-2}, \ldots, \gamma_{ii}$. Computations are conveniently carried out in the following order:

$$\gamma_{11}; \quad \gamma_{22}, \gamma_{12}; \quad \gamma_{33}, \gamma_{23}, \gamma_{13}; \quad \gamma_{44}, \gamma_{34}, \gamma_{24}, \gamma_{14}; \quad \text{etc.}$$

(*d*) *Evaluation of* $(I + UU^{\mathrm{T}})^{-1}$

This inversion is achieved via one more GSO:

(5.1.18) **Theorem:** If U is a $k \times r$ matrix and a GSO is performed on the columns of

$$\begin{matrix} & r \\ k & \\ r & \end{matrix}\begin{pmatrix} U \\ \hline I \end{pmatrix},$$

the resulting matrix of orthonormal vectors is

$$V = \begin{matrix} & r \\ k & \\ r & \end{matrix}\begin{pmatrix} V_1 \\ \hline V_2 \end{pmatrix}$$

where

$$V_2 V_2^{\mathrm{T}} = (I + U^{\mathrm{T}} U)^{-1}$$

and

$$I - V_1 V_1^{\mathrm{T}} = (I + UU^{\mathrm{T}})^{-1}.$$

Comment: The rank of $\binom{U}{I}$ is r since $\binom{U}{I}$ has the same rank as $\binom{U}{I}^{\mathrm{T}}\binom{U}{I} = I + U^{\mathrm{T}}U$ $(r \times r)$ which is nonsingular, (2.13), and hence has rank r. Therefore, the GSO, when performed on the columns of $\binom{U}{I}$ will generate only nonzero vectors.

Proof of theorem: Let $H = \binom{U}{I}$. H has rank r [so that $H^+ H = I$, (3.5.3)] and $\mathscr{R}(V) = \mathscr{R}(H)$, so the equation

$$HZ = V$$

has a solution

$$Z = H^+ V. \tag{3.13.2}$$

Since the columns of V are orthonormal, $V^{\mathrm{T}} V = I$, hence $V^+ = (V^{\mathrm{T}} V)^+ V^{\mathrm{T}} = V^{\mathrm{T}}$. Since $\mathscr{R}(V) = \mathscr{R}(H)$,

$$HH^+ = VV^+ = VV^{\mathrm{T}}.$$

Therefore

$$\begin{aligned} ZZ^{\mathrm{T}} &= H^+(VV^{\mathrm{T}})H^{+\mathrm{T}} = (H^{\mathrm{T}}H)^+ \\ &= (H^{\mathrm{T}}H)^{-1} \qquad \text{since } H \text{ has rank } r. \qquad \text{(4.10.2a) and (3.5.1)} \end{aligned}$$

Since

$$HZ = \left(\frac{U}{I}\right) Z = \left(\frac{V_1}{V_2}\right)$$

we have

$$UZ = V_1$$

and

$$Z = V_2.$$

Therefore

$$V_2 V_2^{\mathrm{T}} = ZZ^{\mathrm{T}} = (H^{\mathrm{T}}H)^{-1} = (I + U^{\mathrm{T}}U)^{-1}.$$

The second part of the theorem follows from the identity

$$(I + UU^{\mathrm{T}})^{-1} = I - U(U^{\mathrm{T}}U + I)^{-1}U^{\mathrm{T}}$$

and the fact that

$$V_1 = UZ = UV_2.$$

Therefore

$$(I + UU^{\mathrm{T}})^{-1} = I - U(V_2 V_2^{\mathrm{T}})\, U^{\mathrm{T}} = I - V_1 V_1^{\mathrm{T}}. \quad \blacksquare$$

Summary of Method I

To find the pseudoinverse of the $n \times m$ matrix A:

1. Perform a GSO on the columns of A. Do not normalize. Call this set of vectors $(c_1^* \mid \cdots \mid c_m^*)$.
2. Permute the c_j^*'s so that $(c_1 \mid c_2 \mid \cdots \mid c_m) = (c_1^* \mid c_2^* \mid \cdots \mid c_m^*)\, P$ where P is a permutation matrix chosen so that

$$c_j \neq 0 \qquad j = 1, 2, \ldots, k$$

$$c_j = 0 \qquad j = k+1, \ldots, m.$$

3. Compute γ_{ij} for $j = i+1, \ldots, k$; $i = 1, \ldots, k$, according to (5.1.13) ($\gamma_{ii} = 1$, $\gamma_{ij} = 0$ if $i < j$).
4. B is the $k \times k$ matrix whose (i,j)th entry is $\gamma_{ij}/\|c_j\|$, U is the $k \times m-k$ matrix whose (i,j)th entry is ω_{ij} [given by (5.1.15)], $(I + UU^{\mathrm{T}})^{-1}$ is obtained by performing a (normalized) GSO on the columns of $\left(\frac{U}{I}\right)$ and

$$Q = \left(\frac{c_1}{\|c_1\|} \,\Big|\, \cdots \,\Big|\, \frac{c_k}{\|c_k\|}\right).$$

5. $A^+ = P(I \mid U)^{\mathrm{T}}(I + UU^{\mathrm{T}})^{-1}BQ^{\mathrm{T}}$.

Comment: A FORTRAN listing for this procedure is given in Rust *et al.* [1].

(5.1.19) **Example**

$$A = \begin{pmatrix} 1 & 0 & 1 & 1 \\ 0 & 1 & -1 & 0 \\ 1 & 1 & 0 & 1 \end{pmatrix},$$

$$c_1 = \begin{pmatrix} 1 \\ 0 \\ 1 \end{pmatrix}, \qquad c_2 = \begin{pmatrix} -\frac{1}{2} \\ 1 \\ \frac{1}{2} \end{pmatrix}, \qquad c_3 = c_4 = \begin{pmatrix} 0 \\ 0 \\ 0 \end{pmatrix},$$

$$k = 2, \qquad P = I, \qquad R = \begin{pmatrix} 1 & 0 \\ 0 & 1 \\ 1 & 1 \end{pmatrix}, \qquad S = \begin{pmatrix} 1 & 1 \\ -1 & 0 \\ 0 & 1 \end{pmatrix}.$$

$$\begin{aligned} \gamma_{11} &= 1 & & \\ \gamma_{22} &= 1 & \gamma_{12} &= -\tfrac{1}{2} \\ \omega_{11} &= 1 & \omega_{12} &= 1 \\ \omega_{21} &= -1 & \omega_{22} &= 0. \end{aligned}$$

$$\begin{aligned} \beta_{11} &= 1/\sqrt{2} & & \\ \beta_{22} &= 2/\sqrt{6} & \beta_{12} &= -1/\sqrt{6}. \end{aligned}$$

$$B = \begin{pmatrix} 1/\sqrt{2} & 0 \\ 2/\sqrt{6} & -1/\sqrt{6} \end{pmatrix}, \qquad Q = \begin{pmatrix} 1/\sqrt{2} & -1/\sqrt{6} \\ 0 & 2/\sqrt{6} \\ 1/\sqrt{2} & 1/\sqrt{6} \end{pmatrix}.$$

$$\begin{pmatrix} U \\ \hline I \end{pmatrix} = \begin{pmatrix} 1 & 0 \\ -1 & 0 \\ \hline 1 & 0 \\ 0 & 1 \end{pmatrix} \xrightarrow{\text{GSO}} V = \begin{bmatrix} 1/\sqrt{3} & 2/\sqrt{15} \\ -1/\sqrt{3} & 1/\sqrt{15} \\ \hline 1/\sqrt{3} & -1/\sqrt{15} \\ 0 & 3/\sqrt{15} \end{bmatrix} = \begin{bmatrix} V_1 \\ \hline V_2 \end{bmatrix}.$$

$$I - V_1 V_1^{\mathrm{T}} = \frac{1}{5}\begin{bmatrix} 2 & 1 \\ 1 & 3 \end{bmatrix} = (I + UU^{\mathrm{T}})^{-1}.$$

$$(I+UU^{\mathrm{T}})^{-1}BQ^{\mathrm{T}} = \frac{1}{15}\begin{pmatrix} 3 & 0 & 3 \\ -1 & 5 & 4 \end{pmatrix}.$$

$$A^{+} = \left(\frac{I}{U^{\mathrm{T}}}\right)(I+UU^{\mathrm{T}})^{-1}BQ^{\mathrm{T}} = \frac{1}{15}\begin{pmatrix} 3 & 0 & 3 \\ -1 & 5 & 4 \\ 4 & -5 & -1 \\ 3 & 0 & 3 \end{pmatrix}.$$

(5.1.20) **Exercise** *Alternate method for inverting* $(I+UU^{\mathrm{T}})$*:* Let the columns of U be denoted by $u_1, u_2, \ldots, u_r$. Let

$$W_0 = I$$

$$W_n = W_{n-1} - \frac{(W_{n-1}u_n)(W_{n-1}u_n)^{\mathrm{T}}}{1+u_n^{\mathrm{T}}W_{n-1}u_n} \qquad n = 1,\ldots,r.$$

Then for each n,

$$W_n = \left(I + \sum_{j=1}^{n} u_j u_j^{\mathrm{T}}\right)^{-1}$$

hence

$$W_r = (I+UU^{\mathrm{T}})^{-1}.$$

[*Hint:* Use (4.6.5) or prove directly by induction on n.]

(5.1.21) **Exercise** *Alternate method for the pseudoinversion of* $(I \mid U)$ (Tewarson [2])

(a) If P is square and $P^{\mathrm{T}}(I \mid U) = S^{\mathrm{T}}$ where $S^{\mathrm{T}}S = I$, then $(I \mid U)^{+} = SP^{\mathrm{T}}$.

(b) If the GSO is applied to the columns of

$$\begin{matrix} & k \\ \begin{matrix} k \\ r \end{matrix} & \left(\frac{I}{U^{\mathrm{T}}}\right) \end{matrix}$$

and the resulting set of orthonormal vectors are denoted by $s_1, s_2, \ldots, s_k$, then there is a $k \times k$ matrix P such that

$$\left(\frac{I}{U^{\mathrm{T}}}\right)P = (s_1 \mid s_2 \mid \cdots \mid s_k).$$

(c) Let $S \equiv (s_1 \mid s_2 \mid \cdots \mid s_k)$ be partitioned:

$$S = \begin{matrix} & k \\ \begin{matrix} k \\ r \end{matrix} & \left(\frac{S_1}{S_2}\right) \end{matrix}.$$

Then

$$P = S_1 \qquad \text{and} \qquad (I \mid U)^{+} = SS_1^{\mathrm{T}}.$$

(5.2) Method II   (Based on Gauss–Jordan elimination, Ben-Israel and Wersan [1], Noble [1])

If A is an $n \times m$ matrix of rank k, there always exists a nonsingular matrix E and an orthogonal matrix P such that

$$(5.2.1) \qquad EA^{\mathsf{T}}AP = \begin{matrix} & k & m-k \\ k & \\ m-k \end{matrix}\left(\begin{array}{c|c} I & L \\ \hline 0 & 0 \end{array}\right).$$

Indeed, there are many such E and P. For instance, if P diagonalizes $A^{\mathsf{T}}A$ in such a way that the nonzero eigenvalues of $A^{\mathsf{T}}A$ appear in the upper left-hand corner:

$$P^{\mathsf{T}}A^{\mathsf{T}}AP = \begin{matrix} & k & m-k \\ k \\ m & k \end{matrix}\left(\begin{array}{c|c} D & 0 \\ \hline 0 & 0 \end{array}\right)$$

then

$$E = \left(\begin{array}{c|c} D^{-1} & 0 \\ \hline 0 & I \end{array}\right)P^{\mathsf{T}}$$

will do the trick:

$$EA^{\mathsf{T}}AP = \left(\begin{array}{c|c} D^{-1} & 0 \\ \hline 0 & I \end{array}\right)(P^{\mathsf{T}}A^{\mathsf{T}}AP) = \left(\begin{array}{c|c} I & 0 \\ \hline 0 & 0 \end{array}\right)$$

which is a stronger version of (5.2.1). However, the method to be described here only requires that E and P be chosen to satisfy (5.2.1). The present method is based upon the following identity:

(5.2.2) Theorem: If E is nonsingular and P is orthogonal, chosen to satisfy (5.2.1), then

$$A^{+} = P(EA^{\mathsf{T}}AP)^{+} EA^{\mathsf{T}}.$$

Proof: The equation

$$A^{\mathsf{T}}AX = A^{\mathsf{T}}$$

always has a solution in X, (3.13.1), (3.13.2), hence the equation

$$(5.2.3) \qquad A^{\mathsf{T}}APY = A^{\mathsf{T}}$$

always has a solution in Y since P is nonsingular, hence

$$(5.2.4) \qquad EA^{\mathsf{T}}APY = EA^{\mathsf{T}}$$

always has a solution in Y.

Among all solutions to (5.2.3), the unique matrix which minimizes $\operatorname{tr}(Y^{\mathsf{T}}Y)$ is

$$(5.2.5) \qquad \hat{Y} = (EA^{\mathsf{T}}AP)^{+} EA^{\mathsf{T}}. \tag{3.13.8}$$

The set of matrices Y satisfying (5.2.4) is the same as the set of matrices satisfying (5.2.3) since E is nonsingular, so $\hat{Y}$ also minimizes

$$\operatorname{tr}(Y^{\mathrm{T}}Y) = \operatorname{tr}[(PY)^{\mathrm{T}}(PY)] \qquad (P \text{ is orthogonal})$$

over that class. By (3.13.8),

$$P\hat{Y} = (A^{\mathrm{T}}A)^{+}A^{\mathrm{T}} = A^{+} \tag{5.2.6}$$

and the theorem follows when (5.2.6) and (5.2.5) are combined. ■

(5.2.7) **Corollary:** If E is nonsingular and P is orthogonal and $EA^{\mathrm{T}}AP = \left(\frac{H}{0}\right)$ then

$$A^{+} = P(H^{+} \mid 0)(EA^{\mathrm{T}}).$$

It remains to show how to compute EA^{T}, H^{+}, and P.

(a) *Computation of H, EA^{T}, and P*

Write down the augmented matrix, ${}_{m}(\overset{m}{A^{\mathrm{T}}A} \mid \overset{n}{A^{\mathrm{T}}})$, and perform *row operations* on this matrix until the left block is reduced to the point where a permutation of its columns will result in the row echelon form:

$$m(\overset{m}{A^{\mathrm{T}}A} \mid \overset{n}{A^{\mathrm{T}}}) \xrightarrow[\text{only (step 1)}]{\text{row operations}} m(\overset{m}{EA^{\mathrm{T}}A} \mid \overset{n}{EA^{\mathrm{T}}})$$

$$\xrightarrow[\substack{m \text{ columns} \\ \text{(step 2)}}]{\text{permute first}} m(\overset{m}{EA^{\mathrm{T}}AP} \mid \overset{n}{EA^{\mathrm{T}}})$$

$$= \begin{matrix} k \\ m-k \end{matrix}\left(\begin{array}{c|c|c} \overset{k}{I} & \overset{m-k}{L} & \\ \hline 0 & 0 & \end{array} EA^{\mathrm{T}}\right).$$

The elementary row operations are reversible so E is nonsingular, as required. The permutation matrix P is orthogonal and is recorded as the columns of $EA^{\mathrm{T}}A$ are permuted to achieve row echelon form. The matrix EA^{T} occupies the right-hand block of the augmented matrix just after Step 1. The matrix H is $k \times m$ and is given by $(I \mid L)$ after Step 2.

(b) *Computation of H^{+}*

$$H^{+} = (I \mid L)^{+} = \left(\frac{I}{L^{\mathrm{T}}}\right)(I + LL^{\mathrm{T}})^{-1}.$$

Evaluate $(I+LL^{\mathrm{T}})^{-1}$ either by (5.1.18) or (5.1.20), or evaluate $(I \mid L)^{+}$ by (5.1.21).

(5.2.8) **Example**

$$A = \begin{bmatrix} -1 & 0 & 1 & 2 \\ -1 & 1 & 0 & -1 \\ 0 & -1 & 1 & 3 \\ 0 & 1 & -1 & -3 \\ 1 & -1 & 0 & 1 \\ 1 & 0 & -1 & -2 \end{bmatrix}$$

$$(A^{\mathrm{T}}A \mid A^{\mathrm{T}}) = \left[\begin{array}{cccc|cccccc} 4 & -2 & -2 & \textcircled{-2} & -1 & -1 & 0 & 0 & 1 & 1 \\ -2 & 4 & -2 & -8 & 0 & 1 & -1 & 1 & -1 & 0 \\ -2 & -2 & 4 & 10 & 1 & 0 & 1 & -1 & 0 & -1 \\ -2 & -8 & 10 & 28 & 2 & -1 & 3 & -3 & 1 & -2 \end{array}\right].$$

Pivot on the 4th element of the first row (which means "reduce the 4th element of all the other rows to zero by subtracting an appropriate multiple of the first row from each"):

$$\to \left[\begin{array}{cccc|cccccc} 4 & -2 & -2 & -2 & -1 & -1 & 0 & 0 & 1 & 1 \\ -18 & 12 & \textcircled{6} & 0 & 4 & 5 & -1 & 1 & -5 & -4 \\ 18 & -12 & -6 & 0 & -4 & -5 & 1 & -1 & 5 & 4 \\ 54 & -36 & -18 & 0 & -12 & -15 & 3 & -3 & 15 & 12 \end{array}\right].$$

Next, pivot on the third element of the second row:

$$\to \left[\begin{array}{cccc|cccccc} -2 & 2 & 0 & -2 & \frac{1}{3} & \frac{2}{3} & -\frac{1}{3} & \frac{1}{3} & -\frac{2}{3} & \frac{1}{3} \\ -18 & 12 & 6 & 0 & 4 & 5 & -1 & 1 & -5 & 4 \\ 0 & 0 & 0 & 0 & 0 & 0 & 0 & 0 & 0 & 0 \\ 0 & 0 & 0 & 0 & 0 & 0 & 0 & 0 & 0 & 0 \end{array}\right].$$

Next, divide the first row by -2, the second by 6:

$$\to \left[\begin{array}{cccc|cccccc} 1 & -1 & 0 & 1 & -\frac{1}{6} & -\frac{2}{6} & \frac{1}{6} & -\frac{1}{6} & \frac{2}{6} & \frac{1}{6} \\ -3 & 2 & 1 & 0 & \frac{4}{6} & \frac{5}{6} & -\frac{1}{6} & \frac{1}{6} & -\frac{5}{6} & -\frac{4}{6} \\ 0 & 0 & 0 & 0 & 0 & 0 & 0 & 0 & 0 & 0 \\ 0 & 0 & 0 & 0 & 0 & 0 & 0 & 0 & 0 & 0 \end{array}\right]$$

$$= (EA^{\mathrm{T}}A \mid EA^{\mathrm{T}}).$$

This completes Step 1; EA^{T} is the right-hand 4×6 block. The left-hand block is reduced to row-echelon form by postmultiplication by

$$P = \begin{bmatrix} 0 & 0 & 1 & 0 \\ 0 & 0 & 0 & 1 \\ 0 & 1 & 0 & 0 \\ 1 & 0 & 0 & 0 \end{bmatrix}$$

which puts column 4 into column 1, column 3 into column 2, column 1 into column 3, and column 2 into column 4. The result:

$$EA^{\mathrm{T}}AP = \begin{pmatrix} H \\ \hline 0 \end{pmatrix} = \left(\begin{array}{cc|cc} 1 & 0 & 1 & -1 \\ 0 & 1 & -3 & 2 \\ \hline & & 0 & \end{array}\right).$$

The matrix L is the upper right-hand block:

$$L = \begin{pmatrix} 1 & -1 \\ -3 & 2 \end{pmatrix}. \qquad I + LL^{\mathrm{T}} = \begin{pmatrix} 3 & -5 \\ -5 & 14 \end{pmatrix}.$$

$$(I+LL^{\mathrm{T}})^{-1} = \frac{1}{17}\begin{pmatrix} 14 & 5 \\ 5 & 3 \end{pmatrix}.$$

$$H^{+} = \begin{bmatrix} (I+LL^{\mathrm{T}})^{-1} \\ \hline L^{\mathrm{T}}(I+LL^{\mathrm{T}})^{-1} \end{bmatrix} = \frac{1}{17}\begin{pmatrix} 14 & 5 \\ 5 & 3 \\ -1 & -4 \\ -4 & 1 \end{pmatrix}.$$

$$(H^{+} \mid 0)\, EA^{\mathrm{T}} = \frac{1}{102}\begin{pmatrix} 6 & -3 & 9 & -9 & 3 & -6 \\ 7 & 5 & 2 & -2 & -5 & -7 \\ -15 & -18 & 3 & -3 & 18 & 15 \\ 8 & 13 & -5 & 5 & -13 & -8 \end{pmatrix}.$$

$$P(H^{+} \mid 0)\, EAT = \frac{1}{102}\begin{pmatrix} -15 & -18 & 3 & -3 & 18 & 15 \\ 8 & 13 & -5 & 5 & -13 & -8 \\ 7 & 5 & 2 & -2 & -5 & -7 \\ -6 & -3 & 9 & -9 & 3 & -6 \end{pmatrix} = A^{+}.$$

(5.2.9) **Exercise:** Let E, A, and P be as defined in (5.2.1). Show that the last $m-k$ rows of EA^{T} are always zero. [*Hint:* $XA^{\mathrm{T}}A = 0$ if and only if $XA^{\mathrm{T}} = 0$. Specialize to matrices of the form

$$X = n(\overset{k}{0} \,\vdots\, \overset{m-k}{Y_1}).]$$

Comment: A GATE 20 program for this procedure is given in Ben-Israel and Ijiri [1]. Additional refinements are contained in Tewarson [1].

(5.3) Method III (Based on gradient projection method, Pyle [1])

If A is an $n \times m$ matrix and $b \in \mathscr{R}(A)$, the equation

(5.3.1) $$Ax = b$$

has at least one solution and

(5.3.2) $$\hat{x} = A^+ b$$

is the only solution which lies in $\mathscr{R}(A^{\mathrm{T}})$, (3.1b).

Let the columns of A^{T} be denoted by $a_1, a_2, \ldots, a_n$ and denote by A_k (resp. b_k) the $k \times m$ matrix (resp. k-vector) obtained by deleting the last $n-k$ rows (resp. components) from A (resp. b). Since (5.3.1) has a solution, so does

(5.3.3) $$A_k x_k = b_k \qquad k = 1, 2, \ldots, n,$$

since the set of simultaneous equations which (5.3.3) represents is a subset of those represented by (5.3.1).

Furthermore,

(5.3.4) $$\hat{x}_k = A_k^{\ +} b_k$$

is the only solution to (5.3.3) which lies in $\mathscr{R}(A_k^{\ \mathrm{T}}) \equiv \mathscr{L}(a_1, \ldots, a_k)$, (3.4, 3.1b). The present procedure develops a simple recursion which relates $\hat{x}_{k+1}$ to $\hat{x}_k$. Since $\hat{x} = A^+ b$ is the same as $\hat{x}_n$, the recursion can be carried out n times and the desired solution to (5.3.1) results, provided b is in $\mathscr{R}(A)$. Extensions to the general case follow readily as we will show:

By definition, $\hat{x}_1$ is the unique vector in $\mathscr{L}(a_1)$ satisfying

$$a_1^{\ \mathrm{T}} x_1 = \beta_1$$

where β_j is the jth component of b ($j = 1, \ldots, n$).

Obviously

(5.3.5) $$\hat{x}_1 = (a_1^{\ \mathrm{T}})^+ \beta_1 = \beta_1 a_1 (\|a_1\|^2)^+.$$

If $\hat{x}_{k-1}$ is known, we construct $\hat{x}_k$ as follows:

Let $(h_1, h_2, \ldots, h_n)$ be the nonnormalized set of vectors which result from a

Gramm–Schmidt orthogonalization of $(a_1, \ldots, a_n)$:

$$h_1 = a_1 \tag{5.3.6}$$

$$h_k = a_k - \sum_{j \in S_k} (a_k^{\mathrm{T}} h_j) h_j / \|h_j\|^2$$

where

$$S_k = \{j : j \leqslant k-1 \quad \text{and} \quad \|h_j\| \neq 0\} \qquad k = 1, \ldots, n.$$

We have shown that $h_k \perp \mathscr{L}(a_1, \ldots, a_{k-1})$, (2.8), and since $\hat{x}_{k-1}$ satisfies

$$A_{k-1} \hat{x}_{k-1} = b_{k-1} \tag{5.3.7}$$

it follows that

$$a_j^{\mathrm{T}}(\hat{x}_{k-1} + \alpha_k h_k) = \beta_j \qquad j = 1, 2, \ldots, k-1 \tag{5.3.8}$$

for any scalar α_k. [Just write (5.3.7) in componentwise form.]

In particular, if

$$\hat{\alpha}_k = \begin{cases} 0 & \text{if} \quad h_k = 0 \\ (\beta_k - a_k^{\mathrm{T}} \hat{x}_{k-1}) / (h_k^{\mathrm{T}} a_k) & \text{otherwise} \end{cases} \tag{5.3.9}$$

then

$$\hat{y}_k = \hat{x}_{k-1} + \hat{\alpha}_k h_k \tag{5.3.10}$$

satisfies

$$a_j^{\mathrm{T}} y = \beta_j \qquad j \in S_{k+1} \tag{5.3.11}$$

and

$$\hat{y}_k \in \mathscr{L}(h_1, \ldots, h_k). \tag{5.3.12}$$

By construction,

$$\begin{aligned} \mathscr{L}(h_j; j \in S_{k+1}) &= \mathscr{L}(h_1, \ldots, h_k) \\ &= \mathscr{L}(a_1, \ldots, a_k) = \mathscr{L}(a_j; j \in S_{k+1}) \end{aligned}$$

since $h_j = 0$ if and only if a_j is a linear combination of its predecessors, (2.8.6). Therefore, $\hat{y}_k$ is the *unique* vector in $\mathscr{L}(a_j; j \in S_{k+1})$ satisfying (5.3.11). On the other hand,

$$\hat{x}_k = A_k^{+} b_k \in \mathscr{L}(a_1, \ldots, a_k) = \mathscr{L}(a_j; j \in S_{k+1})$$

satisfies

$$a_j^{\mathrm{T}} x = \beta_j \qquad j = 1, \ldots, k \tag{5.3.13}$$

and hence it also satisfies (5.3.11). Since $\hat{y}_k$ is the *unique* vector in $\mathscr{L}(a_j; j \in S_{k+1})$ satisfying (5.3.11), $\hat{y}_k = \hat{x}_k$.

In summary:

(5.3.14) **Theorem:** If

$$A^{\mathrm{T}} = (a_1 \,\vdots\, a_2 \,\vdots\, \cdots \,\vdots\, a_n) \qquad \text{and} \qquad b = \begin{pmatrix} \beta_1 \\ \beta_2 \\ \vdots \\ \beta_n \end{pmatrix} \in \mathscr{R}(A)$$

then

$$A^+ b = \hat{x}_n$$

where

$$\hat{x}_0 = 0$$

$$\hat{x}_k = \hat{x}_{k-1} + \hat{\alpha}_k h_k \qquad k = 1, 2, \ldots, n$$

$$\hat{\alpha}_k = \begin{cases} 0 & \text{if} \quad h_k = 0 \\ (\beta_k - a_k{}^{\mathrm{T}} \hat{x}_{k-1})/(h_k{}^{\mathrm{T}} a_k) & \text{otherwise} \end{cases}$$

and $\{h_1, h_2, \ldots, h_n\}$ are the nonnormalized vectors resulting from a GSO of $\{a_1, a_2, \ldots, a_n\}$.

In general, if $b \notin \mathscr{R}(A)$ [or if you are not able to decide whether $b \in \mathscr{R}(A)$ or not, conveniently] resort to the following trick:

Let

$$d = A^{\mathrm{T}} b$$

and

$$C = A^{\mathrm{T}} A.$$

Then $d \in \mathscr{R}(A^{\mathrm{T}}) = \mathscr{R}(C)$ so that (5.3.14) can be used to compute $C^+ d$. Since $C^+ d = (A^{\mathrm{T}} A)^+ A^{\mathrm{T}} b$, this additional step generalizes the procedure to the case of arbitrary b.

If, instead of $A^+ b$, it is desired to compute A^+, proceed in the following manner:

First, perform a GSO on the columns of A. Denote the *normalized* set of nonzero (orthonormal) vectors by $d_1, d_2, \ldots, d_r$. These vectors span the same linear manifold that the columns of A span, namely $\mathscr{R}(A)$. By (3.7.3), it therefore follows that

$$AA^+ = \sum_{j=1}^{r} d_j d_j^{\mathrm{T}}.$$

Denote the columns of AA^+ by $b_1, b_2, \ldots, b_n$. Since $\mathscr{R}(A) = \mathscr{R}(AA^+) = \mathscr{L}(b_1, b_2, \ldots, b_n)$, it must be that $b_j \in \mathscr{R}(A)$ for each j so that (5.3.14) can be used n times to compute $A^+ b_j$ $(j = 1, \ldots, n)$. Since

$$(A^+ b_1 \,\vdots\, A^+ b_2 \,\vdots\, \cdots \,\vdots\, A^+ b_n) = A^+ (b_1 \,\vdots\, b_2 \,\vdots\, \cdots \,\vdots\, b_n) = A^+ (AA^+) = A^+,$$

this procedure generates the desired result.

Summary of Method III

1. If it is known that $b \in \mathscr{R}(A)$ then (5.3.14) can be used directly to construct A^+b.

2. If it is not known whether or not $b \in \mathscr{R}(A)$, compute $d = A^{\mathrm{T}}b$, $C = A^{\mathrm{T}}A$ and use (5.3.14) to compute C^+d which coincides with A^+b.

3. To compute A^+, first perform a GSO on the columns of A. Denote the resulting orthonormal vectors by $d_1, d_2, \ldots, d_r$. Construct the matrix $\sum_{j=1}^{r} d_j d_j^{\mathrm{T}}$ and denote the columns of this matrix by $b_1, b_2, \ldots, b_n$. Then for each j, $b_j \in \mathscr{R}(A)$ and (5.3.14) can be used to construct A^+b_j, $j = 1, \ldots, n$.

$$A^+ = (A^+b_1 \,\vdots\, A^+b_2 \,\vdots\, \cdots \,\vdots\, A^+b_n).$$

(5.3.15) **Exercise:** In (5.3.14) we defined

$$\hat{\alpha}_k = \begin{cases} 0 & \text{if } \ h_k = 0 \\ (\beta_k - a_k^{\mathrm{T}}\hat{x}_{k-1})/(h_k^{\mathrm{T}}a_k) & \text{otherwise.} \end{cases}$$

Show that $h_k^{\mathrm{T}}a_k \neq 0$ if $h_k \neq 0$.

(5.3.16) **Example:** Compute A^+ where A is given by (5.1.19):

$$A = \begin{pmatrix} 1 & 0 & 1 & 1 \\ 0 & 1 & -1 & 0 \\ 1 & 1 & 0 & 1 \end{pmatrix}.$$

The columns of A are

$$c_1 = \begin{pmatrix} 1 \\ 0 \\ 1 \end{pmatrix} \quad c_2 = \begin{pmatrix} 0 \\ 1 \\ 1 \end{pmatrix} \quad c_3 = \begin{pmatrix} 1 \\ -1 \\ 0 \end{pmatrix} \quad c_4 = \begin{pmatrix} 1 \\ 0 \\ 1 \end{pmatrix}.$$

The GSO generates the orthonormal set:

$$d_1 = \begin{pmatrix} 1/\sqrt{2} \\ 0 \\ 1/\sqrt{2} \end{pmatrix} \qquad d_2 = \begin{pmatrix} -1/\sqrt{6} \\ 2/\sqrt{6} \\ 1/\sqrt{6} \end{pmatrix}.$$

$$AA^+ = d_1 d_1^{\mathrm{T}} + d_2 d_2^{\mathrm{T}} = \frac{1}{3}\begin{pmatrix} 2 & -1 & 1 \\ 1 & 2 & 1 \\ 1 & 1 & 2 \end{pmatrix}.$$

The columns of AA^+ are

$$b_1 = \begin{pmatrix} -\frac{2}{3} \\ \frac{1}{3} \\ \frac{1}{3} \end{pmatrix} \qquad b_2 = \begin{pmatrix} -\frac{1}{3} \\ \frac{2}{3} \\ \frac{1}{3} \end{pmatrix} \qquad b_3 = \begin{pmatrix} \frac{1}{3} \\ \frac{1}{3} \\ \frac{2}{3} \end{pmatrix}.$$

To construct A^+b_j, we perform a GSO on the columns of A^{T}: The columns of A^{T} are

$$a_1 = \begin{pmatrix} 1 \\ 0 \\ 1 \\ 1 \end{pmatrix} \qquad a_2 = \begin{pmatrix} 0 \\ 1 \\ -1 \\ 0 \end{pmatrix} \qquad a_3 = \begin{pmatrix} 1 \\ 1 \\ 0 \\ 1 \end{pmatrix}.$$

The nonnormalized GSO produces

$$h_1 = \begin{pmatrix} 1 \\ 0 \\ 1 \\ 1 \end{pmatrix} \qquad h_2 = \begin{pmatrix} \frac{1}{3} \\ 1 \\ -\frac{2}{3} \\ \frac{1}{3} \end{pmatrix} \qquad h_3 = \begin{pmatrix} 0 \\ 0 \\ 0 \\ 0 \end{pmatrix}.$$

To compute A^+b_1, use (5.3.14):

$$\hat{\alpha}_1 = \frac{2\sqrt{3}}{9} \qquad \hat{x}_1 = \frac{2}{9}\begin{pmatrix} 1 \\ 0 \\ 1 \\ 1 \end{pmatrix}$$

$$\hat{\alpha}_2 = \frac{-\sqrt{15}}{45} \qquad \hat{x}_2 = \frac{1}{15}\begin{pmatrix} 3 \\ -1 \\ 4 \\ 3 \end{pmatrix} = A^+b_1.$$

To compute A^+b_2:

$$\hat{\alpha}_1 = \frac{\sqrt{3}}{9} \qquad \hat{x}_1 = \frac{1}{9}\begin{pmatrix} -1 \\ 0 \\ -1 \\ -1 \end{pmatrix}$$

$$\hat{\alpha}_2 = \frac{\sqrt{15}}{9} \qquad \hat{x}_2 = \frac{1}{3}\begin{pmatrix} 0 \\ -1 \\ -1 \\ 0 \end{pmatrix} = A^+ b_2.$$

To compute $A^+ b_3$:

$$\hat{\alpha}_1 = \frac{\sqrt{3}}{9} \qquad \hat{x}_1 = \frac{1}{9}\begin{pmatrix} 1 \\ 0 \\ 1 \\ 1 \end{pmatrix}$$

$$\hat{\alpha}_2 = \frac{4\sqrt{15}}{45} \qquad \hat{x}_2 = \frac{1}{15}\begin{pmatrix} 3 \\ 4 \\ -1 \\ 3 \end{pmatrix} = A^+ b_3.$$

Thus

$$A^+ = (A^+ b_1 \mid A^+ b_2 \mid A^+ b_3) = \frac{1}{15}\begin{pmatrix} 3 & 0 & 3 \\ -1 & 5 & 4 \\ 4 & -5 & -1 \\ 3 & 0 & 3 \end{pmatrix}$$

which agrees with (5.1.19).

(5.4) Method IV (Based on the Cayley–Hamilton theorem, Decell [2], Ben-Israel and Charnes [1])

The present method is based upon two theorems. The first uses the classical Cayley–Hamilton relation (which says that any square matrix satisfies its own characteristic equation) to deduce an expression for the generalized inverse of a matrix in terms of its characteristic polynomial. The second theorem is a result of Fadeev and Fadeeva [1], which generates the coefficients of said characteristic polynomial in an efficient manner.

(5.4.1) **Theorem:** (a) Let A be an $n \times n$ symmetric matrix and let $\pi(\lambda) = \det(A - \lambda I)$, which can always be factored as follows:

$$\pi(\lambda) = \alpha\lambda^k(1 - \lambda\varphi(\lambda))$$

where $n-k$ is the rank of A.

Then if A is nonsingular,

$$A^{-1} = \varphi(A)$$

and generally

$$A^{+} = \varphi(A) + \varphi(0)[A\varphi(A) - I].$$

(b) If H is an arbitrary $m \times n$ matrix and if the characteristic polynomial of $H^{T}H$ is given by $\alpha\lambda^{k}(1-\lambda\varphi(\lambda))$, then

$$H^{+} = \varphi(H^{T}H)\,H^{T}.$$

Comment: The matrix $\varphi(A)$ is obtained by raising A to the appropriate powers and summing these powers using the coefficients of $\varphi(\cdot)$ as weights.

Proof: (a) The Cayley–Hamilton theorem (cf., Bellman [1]) states that A satisfies the matrix equation

$$\pi(A) = 0.$$

Therefore

$$\alpha A^{k}(I - A\varphi(A)) = 0.$$

If A is nonsingular, $k = 0$ and hence

$$A\varphi(A) = I \qquad \text{so that} \qquad \varphi(A) = A^{-1}.$$

Generally,

$$A^{k} = A^{k+1}\varphi(A).$$

Since

$$\begin{aligned}(A^{+})^{k+1}A^{k} &= A^{+}(A^{+k}A^{k}) \\ &= A^{+}(A^{+}A) && (3.11.3) \\ &= A^{+}(AA^{+}) && (3.11.6) \\ &= A^{+} && (3.9.3)\end{aligned}$$

it follows that

$$A^{+} = (A^{+})^{k+1}A^{k+1}\varphi(A).$$

Again, using (3.11.3) and (3.11.6), we see that

(5.4.1.1) $$A^{+} = AA^{+}\varphi(A).$$

Since $\varphi(A) - \varphi(0)$ involves only positive powers of A and since

$$AA^{+}A^{\alpha} = A^{\alpha} \quad \text{if} \quad \alpha > 0 \qquad (3.11.3)$$

it follows from (5.4.1.1) that

$$A^+ = AA^+[\varphi(A)-\varphi(0)] + AA^+\varphi(0)$$

$$(5.4.1.2) \qquad = \varphi(A) - \varphi(0) + AA^+\varphi(0)$$

and so

$$AA^+ = A\varphi(A) + A(I-AA^+)\varphi(0).$$

Since $AA^+ = A^+A$ for symmetric A, (3.11.6), the second term vanishes and

$$(5.4.1.3) \qquad AA^+ = A\varphi(A).$$

Since $\varphi(0)$ is a multiple of the identity, it commutes with $A\varphi(A)$ and so, if (5.4.1.3) is inserted into the right side of (5.4.1.2) we find

$$A^+ = \varphi(A) + (A\varphi(A)-I)\varphi(0)$$

$$= \varphi(A) + \varphi(0)(A\varphi(A)-I).$$

(b) From (a),

$$(H^{\mathrm{T}}H)^+ = \varphi(H^{\mathrm{T}}H) + \varphi(0)[H^{\mathrm{T}}H\varphi(H^{\mathrm{T}}H)-I].$$

From (5.4.1.3) and (3.11.7)

$$H^{\mathrm{T}}H\varphi(H^{\mathrm{T}}H) = H^+H$$

so that

$$H^+ = (H^{\mathrm{T}}H)^+H^{\mathrm{T}} = \varphi(H^{\mathrm{T}}H)H^{\mathrm{T}} + \varphi(0)[H^{\mathrm{T}}H\varphi(H^{\mathrm{T}}H)-I]H^{\mathrm{T}}$$

$$= \varphi(H^{\mathrm{T}}H)H^{\mathrm{T}} + \varphi(0)(H^+H-I)H^{\mathrm{T}}$$

$$= \varphi(H^{\mathrm{T}}H)H^{\mathrm{T}} \qquad (\text{since } H^+HH^{\mathrm{T}} = H^{\mathrm{T}}). \quad \blacksquare$$

Superficially, the last result appears to be of theoretical interest only, owing to the fact that $\varphi(A)$ is defined in terms of the characteristic polynomial of A which, in turn, is defined in terms of a determinant which requires a prohibitively large amount of computation for large matrices. The following result of Fadeev and Fadeeva [1] reduces that problem to manageable proportions:

(5.4.2) **Theorem:** Let A be an $n\times n$ symmetric matrix and define

$$(5.4.2.1) \qquad \left.\begin{cases} A_1 = A \\ \gamma_k = \operatorname{tr} A_k/k \\ B_k = A_k - \gamma_k I \\ A_{k+1} = AB_k \end{cases}\right\} \qquad k = 1,2,\ldots,n-1.$$

($\operatorname{tr} A_k$ is the sum of A_k's diagonal elements.)

Let

(5.4.2.2) $$M = \begin{cases} \text{the first value of } k \leqslant n \text{ for which } AB_k = 0 \text{ if such exists,} \\ n \quad \text{otherwise,} \end{cases}$$

and let

(5.4.2.3) r be the largest $k \leqslant M$ for which $\gamma_k \neq 0$.

Then,

(5.4.2.4) $$\pi(\lambda) \equiv \det(A - \lambda I) = (-1)^n \gamma_r \lambda^{n-r} (1 - \varphi(\lambda))$$

where

(5.4.2.5) $$\varphi(\lambda) = (\gamma_r)^{-1} \left[\lambda^{r-1} - \sum_{j=1}^{r-1} \gamma_j \lambda^{r-1-j} \right].$$

Proof: The proof consists of two parts: First we show that γ_k is the coefficient of

$$(-1)^{n+1} \lambda^{n-k}$$

in the expansion of $\det(A - \lambda I)$ for $k = 1, \ldots, n$. Then we use that result to show that $\gamma_k = 0$ for $k \geqslant r+1$. The desired conclusion follows easily thereafter.

Proof that

(5.4.2.6) $$\det(A - \lambda I) = (-1)^n \left[\lambda^n - \sum_{k=1}^{n} \gamma_k \lambda^{n-k} \right]:$$

Here, $\gamma_1 = \operatorname{tr} A$ which is the coefficient of $(-1)^{n+1} \lambda^{n-1}$ in the expansion of $\det(A - \lambda I)$. We proceed by induction, supposing that $\gamma_1, \gamma_2, \ldots, \gamma_{k-1}$ are, respectively, the coefficients of $(-1)^{n+1} \lambda^{n-1}, (-1)^{n+1} \lambda^{n-2}, \ldots, (-1)^{n+1} \lambda^{n-k+1}$ in the expansion of $\det(A - \lambda I)$.

Define

(5.4.2.7) $$\sigma_k \equiv \operatorname{tr}(A^k).$$

From (5.4.2.1),

$$A_{k+1} = AA_k - \gamma_k A$$

which can be iterated backward to obtain

(5.4.2.8) $$A_k = A^k - \sum_{j=1}^{k-1} \gamma_j A^{k-j} \qquad k = 1, 2, \ldots, n$$

and hence from (5.4.2.7)

(5.4.2.9) $$\operatorname{tr}(A_k) = \sigma_k - \sum_{j=1}^{k-1} \gamma_j \sigma_{k-j}.$$

It is well known that for any A, $\operatorname{tr}(A)$ is the sum of A's eigenvalues, and if $\lambda_1, \lambda_2, \ldots, \lambda_n$ are the eigenvalues of the symmetric matrix A, then $\lambda_1{}^k, \ldots, \lambda_n{}^k$ are the eigenvalues of A^k, so that

$$\sigma_k = \sum_{j=1}^{n} \lambda_j{}^k.$$

Furthermore, Newton's formula states that for any polynomial,

$$\lambda^n - \sum_{j=1}^{n} \beta_j \lambda^{n-j}$$

having roots at $\lambda_1, \lambda_2, \ldots, \lambda_n$, the relations

$$k\beta_k = s_k - \sum_{j=1}^{k-1} \beta_j s_{k-j} \tag{5.4.2.10}$$

hold for $k = 1, 2, \ldots, n$, where

$$s_k = \sum_{j=1}^{k} \lambda_j{}^k \qquad \text{(Bôcher [1])}.$$

Writing

$$\det(A - \lambda I) = (-1)^n \left[\lambda^n - \sum_{j=1}^{n} \beta_j \lambda^{n-j}\right]$$

we see that under the induction hypothesis, $\beta_j = \gamma_j$, $(j = 1, 2, \ldots, k-1)$ so by (5.4.2.10),

$$k\beta_k = \sigma_k - \sum_{j=1}^{k-1} \gamma_j \sigma_{k-j}. \tag{5.4.2.11}$$

By (5.4.2.9) and (5.4.2.1)

$$k\beta_k = \operatorname{tr}(A_k) = k\gamma_k$$

which shows that $\gamma_k = \beta_k$ and hence establishes the induction hypothesis for all k.

The recursion (5.4.2.1) guarantees that $A_k = 0$, $B_k = 0$, and $\gamma_k = 0$ for all $k > M$ if $AB_M = 0$ for some $M < n$. The definition of r therefore implies that $\gamma_k = r$ for $k > r$. From this and (5.4.2.6),

$$\begin{aligned} \det(A - \lambda I) &= (-1)^n \left[\lambda^n - \sum_{j=1}^{r} \gamma_j \lambda^{n-j}\right] \\ &= (-1)^n \gamma_r \lambda^{n-r} \left[1 - (\lambda/\gamma_r)\left(\lambda^{r-1} - \sum_{j=1}^{r-1} \gamma_j \lambda^{r-1-j}\right)\right]. \end{aligned}$$

This proves (5.4.2.4) and (5.4.2.5). ■

The application of (5.4.1) and (5.4.2) to the pseudoinversion of an $n \times m$ matrix is particularly amusing:

(5.4.3) **Theorem:** Let $A = H^T H$ (where H is an arbitrary $m \times n$ matrix), and define A_k, B_k, γ_k, M, and r as in (5.4.2). Then

(5.4.3.1) $$H^+ = (\gamma_r)^{-1} B_{r-1} H^T$$

and the rank of H is r.

Proof: The rank of H is the same as the rank of A. The rank of A is

$$n - (\text{the multiplicity of the characteristic root at } \lambda = 0)$$

since A is symmetric. By (5.4.2.4) this multiplicity is $n-r$, which proves that the rank of H and A is r.

By (5.4.2.5),

$$\varphi(A) = (\gamma_r)^{-1}\left[A^{r-1} - \sum_{j=1}^{r-1} \gamma_j A^{r-1-j}\right]$$

whereas

$$A_{r-1} - q_{r-1} I = A^{r-1} - \sum_{j=1}^{r-1} \gamma_j A^{r-1-j}$$

by (5.4.2.8). Thus

$$\varphi(A) = (\gamma_r)^{-1}(A_{r-1} - \gamma_{r-1} I)$$
$$= (\gamma_r)^{-1} B_{r-1}.$$

The conclusion follows from (5.4.1b). ■

Summary of Method IV

To find H^+ where H is an $m \times n$ matrix:

1. Let $A_1 = H^T H$,
2. $\gamma_k = \operatorname{tr}(A_k/k)$, $B_k = A_k - \gamma_k I$, $A_{k+1} = AB_k$,

$$M = \begin{cases} \text{first value of } k \text{ for which } AB_k = 0 \text{ if such exists and is less than } n \\ n \quad \text{otherwise,} \end{cases}$$

$$r = \text{largest } k \leqslant M \text{ for which } \gamma_k \neq 0,$$

3. $H^+ = (\gamma_r)^{-1} B_{r-1} H^T$.

Comment: Although it is true that $A^{-1} = \varphi(A)$ for nonsingular symmetric A, it is not generally true that $A^+ = \varphi(A)$ as evidenced by the example

$$A = \operatorname{diag}(0, 0, \tfrac{1}{2}, \tfrac{1}{3}).$$

For this example

$$\varphi(\lambda) = 5 - 6\lambda$$

so that

$$\varphi(A) = \text{diag}(5, 5, 2, 3)$$

whereas

$$A^{+} = (0, 0, 2, 3) = \varphi(A) + \varphi(0)[A\varphi(A) - I].$$

(5.4.4) **Example**

$$H = \begin{pmatrix} 1 & 1 & 1 \\ 1 & 0 & -1 \\ 2 & 1 & 0 \\ 3 & 1 & -1 \end{pmatrix} \qquad A_1 = A = H^{\mathrm{T}}H = \begin{pmatrix} 15 & 6 & -3 \\ 6 & 3 & 0 \\ -3 & 0 & 3 \end{pmatrix}$$

$$\gamma_1 = 21$$

$$B_1 = A_1 - 21I = \begin{pmatrix} -6 & 6 & -3 \\ 6 & -18 & 0 \\ -3 & 0 & -18 \end{pmatrix}$$

$$AB_1 = A_2 = 9\begin{pmatrix} -5 & -2 & 1 \\ -2 & -2 & -2 \\ 1 & -2 & -5 \end{pmatrix}$$

$$\gamma_2 = -54$$

$$B_2 = A_2 + 54I = 9\begin{pmatrix} 1 & -2 & 1 \\ -2 & 4 & -2 \\ 1 & -2 & 1 \end{pmatrix}$$

$$A_3 = AB_2 = \begin{pmatrix} 0 & 0 & 0 \\ 0 & 0 & 0 \\ 0 & 0 & 0 \end{pmatrix}$$

$$M = r = 2$$

$$H^{+} = \frac{1}{\gamma_2} B_1 H^{\mathrm{T}} = -\frac{1}{18}\begin{pmatrix} -1 & -1 & -2 & -3 \\ -4 & 2 & -4 & 0 \\ -7 & 5 & -2 & 3 \end{pmatrix}.$$

(5.4.5) **Exercise:** Prove the Cayley–Hamilton theorem for symmetric matrices. (*Hint:* Use the diagonalization theorem.)

(5.4.6) **Exercise:** Establish the result analogous to (5.4.3), using the characteristic polynomial for HH^{T} instead of $H^{\mathrm{T}}H$.

(5.4.7) *General comments*

In Methods I–IV, the option always exists between working with A or A^{T}. For example, in Method I, A^+ can be found by performing a GSO on the columns of A or one can compute $(A^{\mathrm{T}})^+$ by performing a GSO on the columns of A^{T}. Similarly, $(I+UU^{\mathrm{T}})^{-1}$ can be evaluated by performing a GSO on the columns of $\left(\frac{U}{I}\right)$ or on the columns of $\left(\frac{U^{\mathrm{T}}}{I}\right)$. The proper choice is governed mainly by the size of the matrices involved.

All four methods depend crucially upon the computational determination of the rank of the matrix being pseudoinverted. In some of these cases, this is done during the course of a GSO, at that point where one decides whether a vector which is numerically close to zero, is *actually* zero. In Method II, a decision must be made as to whether the Gauss–Jordan elimination has reduced a row to zero or not. If an error (usually due to roundoff) is made at such stages, like as not, the error will be propagated through the remainder of the procedure in a discontinuous manner and cause large errors in the final result. A minor modification significantly improves the GSO in this regard. Whereas the "naive" GSO of (2.8) orthogonalizes the h-vectors in the order they are given (so that for each j, $\mathscr{L}(u_1, \ldots, u_j) = \mathscr{L}(h_1, \ldots, h_j)$, the u's being the orthogonal set induced by the h's), the so called "modified GSO" (Osborne [1]) derives each u_{j+1} from that h-vector whose projection on $\mathscr{L}^{\perp}(u_1, \ldots, u_j)$ is longest. Thus, the borderline cases (is the present h-vector linearly independent of the previously used ones?) are put off until last.

Additional algorithms for pseudoinversion are described in the papers by Pereyre and Rosen [1], Golub and Kahan [1], and Golub [1, 2], all of which pay particular attention to the question of numerical stability.

The recursions of (4.3) and (4.6) can be interpreted as computational algorithms for ${C_m}^+$ and $(C_m {C_m}^{\mathrm{T}})^+$ but the comments above, apply quite strongly in these cases. Extreme care must be exercised in making the decision whether or not a vector c_{m+1} is a linear combination of its predecessors. We will return to these recursions in Chapters VIII and IX.

Part II

STATISTICAL APPLICATIONS

Chapter VI

THE GENERAL LINEAR HYPOTHESIS

In this chapter, we will apply the results obtained so far to some "classical" statistical problems and thereby produce streamlined derivations (and generalizations) for many of the most important results in linear regression analysis. In particular, we will deal with the Gauss–Markov theorem, the distribution theory for best linear unbiased estimator residuals under the normality hypothesis, confidence ellipsoids for linear regression parameters under the normality assumption, the distribution of the likelihood ratio statistic for testing the general linear hypothesis and its relation to confidence ellipsoids, the relationship between Gauss–Markov estimates and "naive" least squares estimates, and the theory of orthogonal designs.

In fairness to the reader, we admit that this chapter presupposes a familiarity with elementary statistical concepts (as exemplified by the usual introductory, but *post calculus*, course in statistics). Such notions as *independence*, *expectation*, *normal distributions*, *tests of hypotheses*, *variance*, etc. are used freely without much in the way of preliminaries.

If things get too heavy, the reader can skip to Chapters VII and VIII which are essentially nonstatistical. Chapter IX is probabilistic in spirit.

(6.1) Best Linear Unbiased Estimation; The Gauss–Markov Theorem

In many applications, observations are made on a random process, and it is felt that a reasonable "model" for the observations is of the form

$$\zeta(\tau_j) = \sum_{i=1}^{p} \alpha_i \varphi_i(\tau_j) + \text{a residual}$$

where τ_j is the value of the jth *independent* variable (e.g., the *time* at which the jth observation was made, or the *concentration* of a catalyst in some chemical experiment, or the *amount* of drug injected into the jth laboratory animal in a drug trial experiment); $\zeta(\tau_j)$ is the *observation* (or *datum*, or *dependent variable*) corresponding to the *level* τ_j; the φ_j's are a specified family of functions and the α's are parameters whose value is to be estimated from the data.

(6.1.1) **Example:** The voltage drop across a legnth of wire is proportional to the current flowing through the wire, the constant of proportionality being the resistance of that wire. It is further known that the resistance of a wire (of uniform thickness) is proportional to its length. The constant of proportionality is the resistance per unit length of the wire and is to be measured as follows. A long length of wire is connected in series to a resistor and a voltage source. See Figure 1. A voltmeter is used to measure the difference

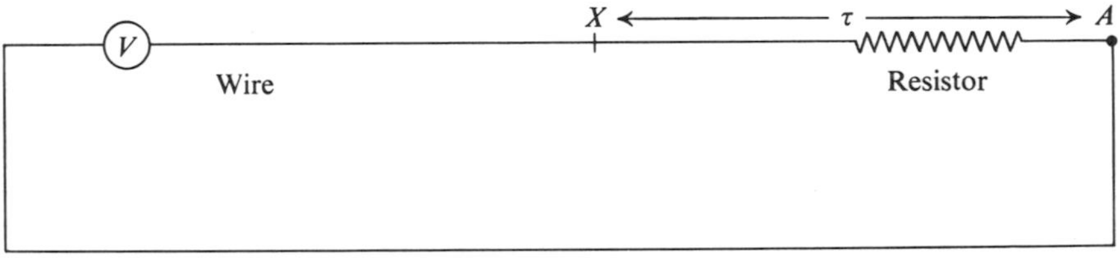

Figure 1.

in voltage between the fixed point A and the variable point X. As X moves, the distance, τ, between X and A changes. The voltage drop, $\zeta(\tau)$, is recorded at various values of τ (say $\tau_1, \tau_2, \ldots, \tau_n$). Since the resistance between A and X is a linear function of τ, a sensible model for this data is

$$\zeta(\tau_j) = \rho\tau_j + R + \text{residual error}$$

so long as X is to the left of the resistor (which is used to prevent "short circuiting" the voltage source). Here, ρ is the resistance per unit length, R is the resistance between A and the left end of the resistor. For this model, $\varphi_1(\tau) = \tau$, $\varphi_2(\tau) = 1$, $\alpha_1 = \rho$, and $\alpha_2 = R$. The residual errors are due to inaccuracy in measurement and in the physical model for the phenomenon being studied.

(6.1.2) **Example:** The mean daily temperature at a given location fluctuates in a quasi-periodic manner with the period of the fluctuation corresponding to the change of seasons. If mean temperatures are recorded daily, then a reasonable model for the temperature $\zeta(\tau_j)$ on the τ_jth day is

$$\zeta(\tau_j) = \alpha_0 + \sum_{i=1}^{k} \alpha_i \cos\frac{2\pi i\tau_j}{365.25} + \beta_i \sin\frac{2\pi i\tau_j}{365.25} + \text{residual}$$

where the period 365.25 takes into account that a calendar year is slightly out of phase with a solar year. The number of harmonics (k) should be chosen reasonably small to keep the model simple. The α's and β's are not known and must be estimated from the data. The residual error now is attributable mainly to the nondeterministic nature of temperature variation rather than errors in measurement.

Both examples are special cases of a so-called *general linear model* in which the vector of observations

$$\mathbf{z} = \begin{pmatrix} \zeta(\tau_1) \\ \zeta(\tau_2) \\ \vdots \\ \zeta(\tau_n) \end{pmatrix} \tag{6.1.3}$$

is assumed to take the form

$$\mathbf{z} = Hx + \mathbf{v} \tag{6.1.4}$$

where H is an $n \times p$ matrix whose jth row is

$$h_j^{\mathrm{T}} = (\varphi_1(\tau_j), \varphi_2(\tau_j), \ldots, \varphi_p(\tau_j)) \tag{6.1.5}$$

x is the p-dimensional vector of unknown coefficients which is to be estimated from the data:

$$x = \begin{pmatrix} \alpha_1 \\ \vdots \\ \alpha_p \end{pmatrix} \tag{6.1.6}$$

and $\mathbf{v}$ is a vector of residuals which is assumed to be a vector random variable with mean value 0:

$$\mathscr{E}\mathbf{v} = 0. \tag{6.1.7}$$

Following Scheffé [1] we define a *parametric function* ψ to be a scalar-valued linear function of the unknown vector parameter x:

$$\psi = c^{\mathrm{T}}x$$

where c is a specified p-dimensional vector. (For instance, each component of x is a *parametric function*.)

A parametric function ψ is said to be *estimable* if it has an *unbiased* linear estimate. (This means that there exists a vector of constant coefficients a, such that

$$\mathscr{E} a^{\mathrm{T}} \mathbf{z} = \psi$$

no matter what the true value of x.) The question of estimability for parametric functions is easily decided:

(6.1.8) **Theorem:** The parametric function $\psi = c^{\mathrm{T}} x$ is estimable if and only if $H^{+} H c = c$ [i.e., if and only if $c \in \mathscr{R}(H^{\mathrm{T}})$].

Proof: Since $\mathscr{E} a^{\mathrm{T}} \mathbf{z} = a^{\mathrm{T}} \mathscr{E} \mathbf{z} = a^{\mathrm{T}} [\mathscr{E} H x + \mathscr{E} \mathbf{v}]$ and since $\mathscr{E} \mathbf{v} = 0$ and Hx is nonrandom, we see that $\psi = c^{\mathrm{T}} x$ is estimable if and only if there exists a vector a such that

(6.1.8.1) $$a^{\mathrm{T}} H x = c^{\mathrm{T}} x \quad \text{for all} \quad x.$$

But (6.1.8.1) holds for all x if and only if there exists an a such that $H^{\mathrm{T}} a - c$ is orthogonal to all x and the last is true if and only if there is a vector a such that

(6.1.8.2) $$H^{\mathrm{T}} a = c. \qquad (2.9.4)$$

The conclusion now follows from (3.12c). ■

The *variance* of an unbiased estimator for ψ is a hallowed measure of its efficacy. Suppose that the residual covariance matrix is denoted by V^2:

(6.1.9) $$\mathscr{E} \mathbf{v} \mathbf{v}^{\mathrm{T}} = V^2$$

where V^2 is an $n \times n$ nonnegative definite matrix. [Since $u^{\mathrm{T}} V^2 u = u^{\mathrm{T}} (\mathscr{E} \mathbf{v} \mathbf{v}^{\mathrm{T}}) u = \mathscr{E} (\mathbf{v}^{\mathrm{T}} u)^2$, it follows that $u^{\mathrm{T}} V^2 u \geqslant 0$ for all u and *is* therefore nonnegative as asserted.]

If $\hat{\boldsymbol{\psi}} = a^{\mathrm{T}} \mathbf{z}$ is an unbiased estimator for ψ, its variance is

(6.1.10) $$\mathscr{E} (\hat{\boldsymbol{\psi}} - \psi)^2 = \mathscr{E} (a^{\mathrm{T}} \mathbf{v})(\mathbf{v}^{\mathrm{T}} a) = a^{\mathrm{T}} V^2 a.$$

The *best* linear unbiased estimator (BLUE) for ψ is (defined to be) the one with minimum variance. It is a quick matter to exhibit a formula for the BLUE of a parametric function:

(6.1.11) **Theorem:** Let $\psi = c^{\mathrm{T}} x$ be an estimable parametric function and let

(6.1.11.1) $$\hat{a} = (I - \bar{V}^{+} V)(H^{+})^{\mathrm{T}} c$$

where

(6.1.11.2) $$\bar{V} = V(I - H H^{+}).$$

Then

(a) $\hat{\psi} = \hat{a}^{\mathrm{T}}\mathbf{z}$ is a BLUE for ψ and
(b) if ψ^* is any other linear unbiased estimator for ψ

$$\mathscr{E}(\psi^* - \psi)^2 > \mathscr{E}(\hat{\psi} - \psi)^2$$

unless $\hat{\psi} = \psi^*$ with probability 1.

Proof: If ψ is estimable, then $a^{\mathrm{T}}\mathbf{z}$ is unbiased if and only if a satisfies (6.1.8.2). By (3.12c), all such a's are of the form

(6.1.11.3) $$a_u = H^{\mathrm{T}+}c - (I - HH^+)u \qquad \text{for some} \quad u.$$

(We used the identity $HH^+ = H^{\mathrm{T}+}H^{\mathrm{T}}$.) The variance of the estimator $\psi_u = a_u^{\mathrm{T}}\mathbf{z}$ is

$$\mathscr{E}(\psi_u - \psi)^2 = \mathscr{E}(a_u^{\mathrm{T}}\mathbf{z} - c^{\mathrm{T}}x)^2 = \mathscr{E}a_u^{\mathrm{T}}(\mathbf{z} - Hx)^2$$

[since a satisfies (6.1.8.2)]

$$= \mathscr{E}(a_u^{\mathrm{T}}\mathbf{v})^2$$

$$= a_u^{\mathrm{T}}V^2 a_u = \|Va_u\|^2$$

(V is the symmetric square root of V^2.) Since

$$\|Va_u\|^2 = \|VH^{\mathrm{T}+}c - \bar{V}u\|^2$$

we see that the variance of the linear unbiased estimator ψ_u is minimized if u is of the form

$$\hat{u} = \bar{V}^+ VH^{\mathrm{T}+}c + (I - \bar{V}^+\bar{V})w$$

where w is free to vary. Thus by (6.1.11.3) and (3.13.10),

$$a_{\hat{u}} = (I - \bar{V}^+ V)(H^+)^{\mathrm{T}}c - (I - HH^+)(I - \bar{V}^+\bar{V})w$$

$$= \hat{a} - (I - HH^+)(I - \bar{V}^+\bar{V})w$$

where w is free to vary. But for such an $a_{\hat{u}}$

$$\mathscr{E}[(a_{\hat{u}} - \hat{a})^{\mathrm{T}}\mathbf{z}]^2 = w^{\mathrm{T}}(I - \bar{V}^+\bar{V})(I - HH^+)V^2(I - HH^+)(I - \bar{V}^+V)w$$

$$= \|\bar{V}(I - \bar{V}^+\bar{V})w\|^2$$

$$= 0 \tag{3.9.1}$$

so that

$$(a_{\hat{u}} - \hat{a})^{\mathrm{T}}\mathbf{z} = 0 \qquad \text{with probability 1.}$$

Thus, *all* BLUE's for ψ are of the form $\hat{a}^{\mathrm{T}}\mathbf{z}$ with probability 1. ■

Comment: The results of (6.1.11) hold whether the residual covariance is singular or not. This result generalizes the Gauss–Markov theorem, which restricts its attention to nonsingular residual covariances (cf. Zyskind and Martin [1], Zyskind [1]).

Theorem (6.1.11) yields the classical result as a special case:

(6.1.12) **Theorem** (Gauss–Markov): Let

$$\bar{V} = V(I - HH^+),$$

(6.1.12.1)
$$G = H^+[I - (\bar{V}^+ V)^T]$$

and

(6.1.12.2)
$$\hat{\mathbf{x}} = G\mathbf{z}.$$

Then

(a) $\mathscr{E}\hat{\mathbf{x}} = H^+Hx$ for all x.

(b) If $\psi = c^T x$ is an estimable parametric function, there is a unique BLUE for ψ, namely $c^T\hat{\mathbf{x}}$.

(c) If V^2 is nonsingular

$$\hat{\mathbf{x}} = (H^T V^{-2} H)^+ H^T V^{-2} \mathbf{z}.$$

Proof: (a) $\mathscr{E}\hat{\mathbf{x}} = GHx = H^+Hx - H^+V^T(\bar{V}^+)^T Hx.$

But

$$(\bar{V}^+)^T H = (\bar{V}\bar{V}^T)^+ \bar{V}H \tag{3.8.1}$$

and

$$\bar{V}H = 0. \tag{6.11.2}$$

(b) The BLUE for ψ is unique by (6.1.11), and is expressible in the form

$$\hat{\psi} = \hat{a}^T\mathbf{z}$$

where

$$\hat{a} = [H^+(I - (\bar{V}^+ V)^T)]^T c = G^T c.$$

Thus

$$\hat{\psi} = c^T\hat{\mathbf{x}} \qquad \text{as asserted.}$$

(c) By (6.1.12.1),

$$G = H^+V[I - (\bar{V}^{+T}V)]V^{-1}$$

if V is nonsingular.

By (3.13.10)

$$\bar{V}^{+\mathrm{T}} = [(I-HH^{+})\bar{V}^{+}]^{\mathrm{T}}$$

so

$$\bar{V}^{+\mathrm{T}}V = \bar{V}^{+\mathrm{T}}\bar{V}^{\mathrm{T}}$$

hence

$$G = H^{+}V[I-\bar{V}^{\mathrm{T}+}\bar{V}^{\mathrm{T}}]V^{-1}.$$

By (4.15) the last is exactly

$$(V^{-1}H)^{+}V^{-1}$$

since

$$\bar{V}^{\mathrm{T}} = QV \qquad \text{where} \quad Q = I - HH^{+}$$

so that

$$G = (V^{-1}H)^{+}V^{-1} = (H^{\mathrm{T}}V^{-2}H)^{+}H^{\mathrm{T}}V^{-2} \tag{3.8.1}$$

and

$$\hat{\mathbf{x}} = G\mathbf{z} = (H^{\mathrm{T}}V^{-2}H)^{+}H^{\mathrm{T}}V^{-2}\mathbf{z}. \quad \blacksquare$$

Comment: When V^2 is nonsingular, then $\hat{\mathbf{x}} = (V^{-1}H)^{+}V^{-1}\mathbf{z}$ minimizes

$$\|V^{-1}\mathbf{z}-V^{-1}Hx\|^2 = (\mathbf{z}-Hx)^{\mathrm{T}}V^{-2}(\mathbf{z}-Hx)$$

which is a weighted sum of squares. In the special case where

$$V = \operatorname{diag}(v_1, v_2, \ldots, v_n)$$

$$\mathbf{z} = \begin{pmatrix} \zeta_1 \\ \vdots \\ \zeta_n \end{pmatrix} \qquad \text{and} \qquad H = \begin{pmatrix} h_1^{\mathrm{T}} \\ \hline h_2^{\mathrm{T}} \\ \hline \vdots \\ \hline h_n^{\mathrm{T}} \end{pmatrix}$$

$\hat{\mathbf{x}}$ minimizes

$$\sum_{=1}^{n} (\zeta_j - h_j^{\mathrm{T}}x)/v_j^{2}$$

which is the sum of squares of the deviations between each observation, ζ_j, and its expectation, each term being weighted according to that observation's accuracy (v_j^2 = variance of ζ_j). Accurate observations (small v_j) are weighted more heavily than inaccurate ones. This interpretation is explored in greater depth in Chapter VII, where we examine the relationship between constrained least squares estimators and weighted least squares.

(6.1.13) **Counterexample:** If V^2 is singular, the BLUE for $c^{\mathrm{T}}x$ is not $c^{\mathrm{T}}(H^{\mathrm{T}}V^{+2}H)^{+}H^{\mathrm{T}}V^{+2}\mathbf{z}$: Take

$$V = \begin{pmatrix} 1 & 0 \\ 0 & 0 \end{pmatrix} \qquad H = \begin{pmatrix} 1 \\ 1 \end{pmatrix} \qquad x = \xi \quad \text{(scalar)} \qquad \mathbf{z} = \begin{pmatrix} \zeta_1 \\ \zeta_2 \end{pmatrix}.$$

Componentwise,

$$\zeta_1 = \xi + \mathbf{v}_1 \qquad \text{where } \mathbf{v}_1 \text{ has unit variance}$$

$$\zeta_2 = \xi.$$

The second observation gives an error-free look at the unknown parameter and the best estimator would be ζ_2.

On the other hand,

$$(H^{\mathrm{T}}V^{+2}H)^{+} = 1 \qquad H^{\mathrm{T}}V^{+2} = (1 \quad 0)$$

so if $c = 1$,

$$c^{\mathrm{T}}(H^{\mathrm{T}}V^{+2}H)^{+}H^{\mathrm{T}}V^{+2}\mathbf{z} = \zeta_1$$

which is not correct.

The correct result is obtained if (6.1.12.2) is used:

$$H^{+} = (\tfrac{1}{2} \quad \tfrac{1}{2}) \qquad I - HH^{+} = \begin{pmatrix} \frac{1}{2} & -\frac{1}{2} \\ -\frac{1}{2} & \frac{1}{2} \end{pmatrix} \qquad \overline{V} = \begin{pmatrix} \frac{1}{2} & -\frac{1}{2} \\ 0 & 0 \end{pmatrix}$$

$$\overline{V}^{+} = \begin{pmatrix} 1 & 0 \\ -1 & 0 \end{pmatrix} \qquad \overline{V}^{+}V = \begin{pmatrix} 1 & 0 \\ -1 & 0 \end{pmatrix};$$

$$I - (\overline{V}^{+}V)^{\mathrm{T}} = \begin{pmatrix} 0 & 1 \\ 0 & 1 \end{pmatrix} \qquad G = H^{+}[I - (\overline{V}^{+}V)^{\mathrm{T}}] = (0 \quad 1) \qquad G\mathbf{z} = \zeta_2.$$

(6.1.14) **Exercise:** When do least squares estimators coincide with Gauss–Markov estimators?

If $\mathbf{z} = Hx + \mathbf{v}$ where $\mathscr{E}\mathbf{v}\mathbf{v}^{\mathrm{T}} = V^2$, the *naive least squares* estimate is defined to be

$$\hat{\hat{\mathbf{x}}} = H^{+}\mathbf{z}.$$

If $\psi = c^{\mathrm{T}}x$ is an estimable parametric function, the naive least squares estimate for ψ is defined to be $c^{\mathrm{T}}\hat{\hat{\mathbf{x}}}$.

(6.1.14.1) Show that the naive least squares estimate for every estimable parametric function is a BLUE if and only if

$$\overline{V}^{+}\overline{V} = (\overline{V}^{+}V)$$

where

$$\bar{V} = V(I - HH^{+}).$$

(6.1.14.2) If V is nonsingular, the condition in (6.1.14.1) reduces to

$$(\bar{V})^{+} = (I - HH^{+})V^{-1}.$$

(6.1.14.3) In general, the condition (6.1.14.1) is equivalent to

$$\mathscr{R}(V^2 H) \subseteq \mathscr{R}(H)$$

(i.e., the range of H is invariant under V^2).

Comment (Jocular): Naive least squares estimates $\hat{\hat{\mathbf{x}}}$ are chosen so that $\|\mathbf{z} - Hx\|^2$ has "tiniest residual." If $c^{\mathrm{T}}x$ is estimable, $c^{\mathrm{T}}\hat{\hat{\mathbf{x}}}$ is unbiased. Therefore, $c^{\mathrm{T}}\hat{\hat{\mathbf{x}}}$ is a TRUE (tiniest residual unbiased estimator) for $c^{\mathrm{T}}x$. (6.7.3) says that $\hat{\hat{\mathbf{x}}}$ is TRUE-BLUE if and only if $\mathscr{R}(V^2 H) \subseteq \mathscr{R}(H)$.

Other conditions equivalent to (6.7) are given in Zyskind [1], Krusbal [1], Mitra *et al.* [1], and Watson [1]. For instance, it is shown by Zyskind that (6.7.1) is equivalent to $\mathscr{R}(H) = \mathscr{L}(y_1, \ldots, y_r)$ where the y_j's are a subset of V's eigenvectors. Mitra *et al.* show that another necessary and sufficient condition is that V^2 should be of the form

$$V^2 = HSH^{\mathrm{T}} + (I - HH^{+})T(I - HH^{+}) + k^2 I$$

where S, T are nonnegative definite and k is a scalar.

(6.1.15) **Exercise** *Generalized least squares* (Price [1])

It has been shown that $\hat{x} = H^{+}z$ is the unique vector which minimizes $\|x\|^2 \equiv x^{\mathrm{T}}x$, among those which also minimize $(z - Hx)^{\mathrm{T}}(z - Hx)$. Let V^2 and W^2 be nonnegative-definite matrices.

Show that among those x's which minimize $(z - Hx)^{\mathrm{T}}V^2(z - Hx)$, one which minimizes $\|x\|_{W}^{2} \equiv x^{\mathrm{T}}W^2 x$ is

$$\bar{x} = [I - (PW^2P)^{+}PW^2](H^{\mathrm{T}}V^2H)^{+}H^{\mathrm{T}}V^2 z$$

where

$$P = I - (VH)^{+}(VH).$$

Are there others?

(6.2) Distribution for Quadratic Forms in Normal Random Variables

In this section, we will investigate the distribution theory for certain types of quadratic forms in random variables which have normal distributions and apply these results to regression analysis (tests of the general linear hypothesis and construction of confidence ellipsoids).

We first review some well-known facts about normal random variables (Scheffé [1]; Anderson [1]).

(6.2.1) If $\xi_1, \xi_2, \ldots, \xi_n$ have a joint normal distribution, they are mutually independent if and only if their covariance matrix is diagonal.

(6.2.2) If $\xi_1, \xi_2, \ldots, \xi_n$ are independent and normally distributed with means $\mu_1, \mu_2, \ldots, \mu_n$ and unit variances, then the distribution of

$$\sum_{=1}^{n} \xi_j^2$$

depends on the μ_j's only through

$$\delta = \left(\sum_{j=1}^{n} \mu_j^2\right)^{1/2}$$

and this distribution is a *noncentral chi-squared* distribution with n degrees of freedom and noncentrality parameter δ. If $\delta = 0$, the distribution is a chi-square distribution (central), with reference to centrality usually omitted.

(6.2.3) If χ_n^2 has a noncentral chi-square distribution with n degrees of freedom and noncentrality parameter δ and if χ_m^2 is independent of χ_n^2 and has a (central) chi-square distribution with m degrees of freedom, then

$$n^{-1}\chi_n^2/m^{-1}\chi_m^2$$

has a *noncentral F* distribution with n and m degrees of freedom and noncentrality parameter δ. Reference to centrality is usually omitted when $\delta = 0$.

(6.2.4) The random vector $\mathbf{z}$ has a multivariate normal distribution with mean m and covariance R if and only if there exists a matrix H such that

(a) $\mathbf{z} = H\mathbf{x}+m$, where the components of $\mathbf{x}$ are independent normal random variables with mean zero and unit variance, and

(b) $R = HH^{\mathrm{T}}$.

If $\left(\frac{\mathbf{u}}{\mathbf{v}}\right)$ has a zero mean multivariate normal distribution, then $\mathbf{u}$ is independent of $\mathbf{v}$ if and only if

$$\mathscr{E}\mathbf{u}\mathbf{v}^{\mathrm{T}} = 0.$$

If $\mathbf{z}$ is a vector random variable which has a multivariate normal distribution with mean m and nonsingular covariance R, the density function of $\mathbf{z}$ has the form

$$\text{const} \times \exp - \tfrac{1}{2}(z-m)^{\mathrm{T}}R^{-1}(z-m).$$

There is a well known proverb in statistical lore that asserts "the exponent of a normal distribution with nonsingular covariance has a chi-square distribution." We will now prove a generalized version of this statement.

(6.2.5) **Theorem:** If $\mathbf{z}$ is a vector random variable having a normal distribution with mean m and covariance R and if the rank of R is ρ, then

(a) $\mathbf{z}^{\mathrm{T}}R^{+}\mathbf{z}$ has a noncentral chi-square distribution with ρ degrees of freedom (df) and noncentrality parameter

$$\delta = (m^{\mathrm{T}}R^{+}m)^{1/2}.$$

(b) $(\mathbf{z}-m)^{\mathrm{T}}R^{+}(\mathbf{z}-m)$ has a chi-square distribution with ρ df.

Proof: R is nonnegative definite since

$$u^{\mathrm{T}}Ru = \mathscr{E}(u^{\mathrm{T}}\mathbf{x})^2 \geqslant 0 \qquad \text{where} \quad \mathbf{x} = \mathbf{z} - m,$$

so R has a symmetric square root $R^{1/2}$. Let $m_1 = (R^{1/2})^{+}m$ and $m_2 = (I - R^{1/2}R^{1/2+})m$. Then

$$m = R^{1/2}m_1 + m_2.$$

If $\mathbf{y}$ is a normally distributed random vector with mean zero and covariance I, then

$$\bar{\mathbf{z}} \equiv R^{1/2}(\mathbf{y}+m_1) + m_2$$

has the same distribution as $\mathbf{z}$, hence $\bar{\mathbf{z}}^{\mathrm{T}}R^{+}\bar{\mathbf{z}}$ has the same distribution as $\mathbf{z}^{\mathrm{T}}R^{+}\mathbf{z}$ and it will suffice to establish (a) for $\bar{\mathbf{z}}^{\mathrm{T}}R^{+}\bar{\mathbf{z}}$:

$$\begin{aligned}\bar{\mathbf{z}}^{\mathrm{T}}R^{+}\bar{\mathbf{z}} &= (\mathbf{y}+m_1)^{\mathrm{T}}R^{1/2}R^{+}R^{1/2}(\mathbf{y}+m_1)\\ &\quad + 2(\mathbf{y}+m_1)^{\mathrm{T}}R^{1/2}R^{+}m_2\\ &\quad + m_2^{\mathrm{T}}R^{+}m_2.\end{aligned}$$

Since $R^{1/2}R^{1/2+} = RR^{+} = R^{+}R$, it is easy to see that $R^{+}m_2 = 0$. Using the diagonalization theorem, it is easy to see that

$$R^{1/2}R^{+}R^{1/2} = RR^{+} = \sum_{i=1}^{\rho} r_i r_i^{\mathrm{T}}$$

where the r_i's are the orthonormal eigenvectors associated with the nonzero eigenvalues of R:

$$R = \sum_{j=1}^{\rho} \lambda_j r_j r_j^{\mathrm{T}} \qquad R^{1/2} = \sum_{j=1}^{\rho} \lambda_j^{1/2} r_j r_j^{\mathrm{T}}$$

$$R^{+} = \sum_{j=1}^{\rho} \lambda_j^{-1} r_j r_j^{\mathrm{T}}. \qquad \text{[cf. (3.6)]}$$

$$RR^{+} = R^{+}R = \sum_{j=1}^{\rho} r_j r_j^{\mathrm{T}}.$$

Thus

$$\bar{\mathbf{z}}^{\mathrm{T}} R^{+} \bar{\mathbf{z}} = (\mathbf{y}+m_1)^{\mathrm{T}} RR^{+}(\mathbf{y}+m_1)$$

$$= \sum_{j=1}^{\rho} [r_j{}^{\mathrm{T}}(\mathbf{y}+m_1)]^2.$$

The vector random variable

$$P(\mathbf{y}+m_1)$$

where

$$P = \begin{pmatrix} r_1{}^{\mathrm{T}} \\ \hline \vdots \\ \hline r_\rho{}^{\mathrm{T}} \end{pmatrix}$$

has a multivariate normal distribution with mean Pm_1 and covariance

$$P(\mathscr{E}\mathbf{y}\mathbf{y}^{\mathrm{T}})\, P^{\mathrm{T}} = PP^{\mathrm{T}} \qquad \text{since} \quad \mathscr{E}\mathbf{y}\mathbf{y}^{\mathrm{T}} = I.$$

The rows of P are orthonormal, so $PP^{\mathrm{T}} = I$ (of order $\rho \times \rho$), hence by (6.2.1), the components of $P(\mathbf{y}+m_1)$ are *independent* normal with mean zero and unit variance, so by (6.2.2)

$$\|P(\mathbf{y}+m_1)\|^2$$

has a noncentral chi-square distribution with ρ degrees of freedom and noncentrality parameter $\|Pm_1\|$. But

$$\|P(\mathbf{y}+m_1)\|^2 = (\mathbf{y}+m)^{\mathrm{T}} P^{\mathrm{T}} P(\mathbf{y}+m)$$

and

$$P^{\mathrm{T}} P = \sum_{j=1}^{\rho} r_j r_j{}^{\mathrm{T}} = R^{+} R.$$

Therefore

$$\|P(\mathbf{y}+m_1)\|^2 = \bar{\mathbf{z}}^{\mathrm{T}} R^{+} \bar{\mathbf{z}}^{\mathrm{T}}$$

and

$$\|Pm_1\|^2 = m_1{}^{\mathrm{T}} RR^{+} m_1 = m^{\mathrm{T}} R^{+} m = \delta^2$$

since

$$m_1{}^{\mathrm{T}} RR^{+} m_1 = \|RR^{+} m_1\|^2$$

and

$$RR^{+} m_1 = RR^{+}(R^{1/2})^{+} m = R^{+1/2} m.$$

Part (b) follows as a special case of part (a) when $m = 0$, since $\bar{\mathbf{z}} - m$ has mean zero. ■ (See Rao [1], for related results.)

(6.2.6) **Exercise:** If $\mathbf{z}$ has a multivariate distribution with mean m and covariance $\sigma^2 I$, and if R is a projection matrix with rank ρ, then

$$\sigma^{-2}\,\|R\mathbf{z}\|^2$$

has a noncentral chi–square distribution with noncentrality parameter

$$\sigma^{-1}(m^{\mathrm{T}}Rm)^{1/2} \qquad \text{and} \qquad \rho \text{ df.}$$

(6.3) Estimable Vector Parametric Functions and Confidence Ellipsoids in the Case of Normal Residuals

In (6.1) we introduced the notion of an estimable parametric function (a real-valued linear function of the unknown vector parameter). We now extend this notion to vector-valued linear functions:

If

$$\mathbf{z} = Hx + \mathbf{v}$$

where $\mathbf{v}$ is a zero mean vector random variable, H is a known matrix and x is an unknown vector parameter, we define an *estimable vector parametric function* (evpf) to be any vector-valued linear function of x whose components are estimable parametric functions. Thus, an evpf is any vector of the form $y = Gx$ where G is a matrix whose rows are $g_1^{\mathrm{T}}, g_2^{\mathrm{T}}, \ldots, g_k^{\mathrm{T}}$, where, for each j, $g_j^{\mathrm{T}}x$ is an estimable parametric function.

It is a trivial consequence of (6.1.8) that

(6.3.1) Gx is an evpf if and only if $\mathscr{R}(G^{\mathrm{T}}) \subseteq \mathscr{R}(H^{\mathrm{T}})$.

By (3.13.1), the last statement is equivalent to

(6.3.2) Gx is an evpf if and only if $GH^{+}H = G$.

If

$$G = \begin{pmatrix} g_1^{\mathrm{T}} \\ \vdots \\ g_k^{\mathrm{T}} \end{pmatrix}$$

is an evpf, then each component of Gx has a BLUE, namely $g_j^{\mathrm{T}}\hat{\mathbf{x}}$ where $\hat{\mathbf{x}}$ is the Gauss–Markov estimator defined in (6.1.12). Henceforth, we will call $G\hat{\mathbf{x}}$ the BLUE for the evpf Gx.

For the sake of the following discussion, let us focus our attention on the case where the residuals $\mathbf{v}$ are normally distributed and have covariance V^2 proportional to the identity matrix:

(6.3.3) $$V^2 = \mathscr{E}\mathbf{v}\mathbf{v}^{\mathrm{T}} = \sigma^2 I.$$

According to (6.1.12), the BLUE for an evpf, $y = Gx$, is $\hat{\mathbf{y}} = G\hat{\mathbf{x}}$, where

$$\hat{\mathbf{x}} = H^{+}\mathbf{z}$$

happens to be the least squares estimate for x.

The vector random variable $\hat{\mathbf{y}}-y$ has zero mean (since $G\hat{\mathbf{x}}$ is unbiased), and the covariance of $\hat{\mathbf{y}}-y$ is

$$\mathscr{E}(\hat{\mathbf{y}}-y)(\hat{\mathbf{y}}-y)^{\mathrm{T}}.$$

But

$$y = Gx = GH^{+}Hx \tag{6.3.2}$$

and

$$\hat{\mathbf{y}} = G\hat{\mathbf{x}} = GH^{+}\mathbf{z} = GH^{+}Hx + GH^{+}\mathbf{v}$$

so that

$$\hat{\mathbf{y}} - y = GH^{+}\mathbf{v}$$

and

(6.3.4) $$\operatorname{Cov}(\hat{\mathbf{y}}-y) \equiv \mathscr{E}(\hat{\mathbf{y}}-y)(\hat{\mathbf{y}}-y)^{\mathrm{T}} = \mathscr{E}[(GH^{+})\mathbf{v}\mathbf{v}^{\mathrm{T}}(GH^{+})^{\mathrm{T}}]$$
$$= \sigma^2 G(H^{\mathrm{T}}H)^{+}G^{\mathrm{T}}.$$

Therefore, $\hat{\mathbf{y}}-y$ has a multivariate normal distribution with mean zero and covariance $\sigma^2 G(H^{\mathrm{T}}H)^{+}G^{\mathrm{T}}$ so, by virtue of (6.2.5),

(6.3.5) $$\sigma^{-2}(y-\hat{\mathbf{y}})^{\mathrm{T}}[G(H^{\mathrm{T}}H)^{+}G^{\mathrm{T}}]^{+}(y-\hat{\mathbf{y}})$$

has a (central) chi-square distribution with k_1 $(=\operatorname{rank} GH^{+})$ degrees of freedom, since

$$\operatorname{rk}(A) = \operatorname{rk}(AA^{\mathrm{T}}) \qquad \text{for any matrix} \quad A. \tag{4.10.2}$$

It is always true that

(6.3.6) $$\mathscr{R}(GH^{+}H) \subseteq \mathscr{R}(GH^{+}) \subseteq \mathscr{R}(G)$$

and if G is an evpf

$$\mathscr{R}(G) = \mathscr{R}(GH^{+}H) \quad \text{as well} \tag{6.3.2}$$

so in this case,

(6.3.6.1) $$\mathscr{R}(GH^{+}) = \mathscr{R}(G).$$

Therefore

$$k_1 = \operatorname{rk}(G).$$

The last result [concerning the distribution of (6.3.5)] can be used to construct a confidence set for y provided σ^2 is *known*. For if $\omega_{1-\alpha}$ is the upper

$100(1-\alpha)$ percentile of said chi-square distribution, then

$$\Pr\{(y-\hat{\mathbf{y}})^{\mathrm{T}}[G(H^{\mathrm{T}}H)^{+}G^{\mathrm{T}}]^{+}(y-\hat{\mathbf{y}}) \leqslant \sigma^2\omega_{1-\alpha}\} = 1-\alpha$$

which means that the ellipsoid

$$\mathscr{S}(\sigma^2\omega_{1-\alpha}) = \{u\colon (u-\hat{\mathbf{y}})^{\mathrm{T}}[G(H^{\mathrm{T}}H)^{+}G^{\mathrm{T}}]^{+}(u-\hat{\mathbf{y}}) \leqslant \sigma^2\omega_{1-\alpha}\}$$

covers the true value of y with probability $1-\alpha$.

If σ^2 is *not known*, its maximum likelihood estimate is proportional to

(6.3.7) $$\hat{\boldsymbol{\sigma}}^2 = (n-k_2)^{-1}\,\|\mathbf{z}-H\hat{\mathbf{x}}\|^2$$

where n is the number of observations ($=$number of rows in H) and k_2 is the rank of H.

We will now show that

(6.3.8) $$\begin{cases} \sigma^{-2}\,\|\mathbf{z}-H\hat{\mathbf{x}}\|^2 \text{ has a chi-square distribution} \\ \text{with } n-k_2 \text{ df and is independent of } y-\hat{\mathbf{y}}. \end{cases}$$

Once (6.3.8) is established, then from (6.2.3)

(6.3.9) $$\frac{(k_1\sigma^2)^{-1}(y-\hat{\mathbf{y}})[G(H^{\mathrm{T}}H)^{+}G^{\mathrm{T}}]^{+}(y-\hat{\mathbf{y}})}{\sigma^{-2}\hat{\boldsymbol{\sigma}}^2}$$

$$= (k_1\hat{\boldsymbol{\sigma}}^2)^{-1}(y-\hat{\mathbf{y}})^{\mathrm{T}}[G(H^{\mathrm{T}}H)^{+}G^{\mathrm{T}}]^{+}(y-\hat{\mathbf{y}})$$

has an F distribution with k_1 and $n-k_2$ df. Because of this, the ellipsoid $\mathscr{S}(k_1\hat{\boldsymbol{\sigma}}^2\tilde{\omega}_{1-\alpha})$ covers the true value of y with probability $1-\alpha$ if $\tilde{\omega}_{1-\alpha}$ is the upper $100(1-\alpha)$ percentile of said F distribution, since

$$\Pr\{y \in \mathscr{S}(k_1\hat{\boldsymbol{\sigma}}^2\tilde{\omega}_{1-\alpha})\}$$
$$= \Pr\{(k_1\hat{\boldsymbol{\sigma}}^2)^{-1}(y-\hat{\mathbf{y}})^{\mathrm{T}}[G(H^{\mathrm{T}}H)^{+}G^{\mathrm{T}}]^{+}(y-\hat{\mathbf{y}}) \leqslant \tilde{\omega}_{1-\alpha}\} = 1-\alpha.$$

To prove (6.3.8), we notice that

$$\sigma^{-1}(\mathbf{z}-H\hat{\mathbf{x}}) = \sigma^{-1}(I-HH^{+})\mathbf{z}$$
$$= \sigma^{-1}(I-HH^{+})(Hx+\mathbf{v})$$
(6.3.10) $$= \sigma^{-1}(I-HH^{+})\mathbf{v}$$

where $\sigma^{-1}\mathbf{v}$ has a multivariate normal distribution with mean zero and covariance I. Furthermore,

$$\hat{\mathbf{y}} - y = GH^{+}\mathbf{v} \tag{6.3.4}$$

hence

$$\sigma^{-2}\mathscr{E}(\hat{\mathbf{y}}-y)(\mathbf{z}-H\hat{\mathbf{x}})^{\mathrm{T}} = GH^{+}(I-HH^{+}) = 0$$

so that $(\hat{\mathbf{y}}-y)$ and $\mathbf{z}-H\hat{\mathbf{x}}$ are jointly normally distributed (both are linear functions of $\mathbf{v}$) and are uncorrelated. By (6.2.4) they must be independent. So, then, are

$$\hat{\boldsymbol{\sigma}}^2 = (n-k_2)^{-1}\|\mathbf{z}-H\hat{\mathbf{x}}\|^2 \qquad \text{and} \qquad (y-\hat{\mathbf{y}}).$$

(6.3.10) shows that $\sigma^{-1}(\mathbf{z}-H\hat{\mathbf{x}})$ is of the form $\sigma^{-1}P\mathbf{v}$ where P is a projection, so by (6.2.6) $(\sigma^2)^{-1}\|\mathbf{z}-H\hat{\mathbf{x}}\|^2$ has a chi-square distribution with as many degrees of freedom as the rank of $I-HH^+$.

(6.3.11) **Exercise**

$$\operatorname{rk}(I-HH^+) = \operatorname{rk}(I) - \operatorname{rk}(H).$$

[*Hint:* I is a projection and so is HH^+. Use (4.10.1)–(4.10.3).]

(6.3.12) **Exercise:** Generalize the results of (6.3) to the case where the residuals have a normal distribution with zero mean and covariance $\sigma^2 V^2$ where σ^2 is unknown and V^2 is a known nonnegative-definite matrix.

(6.4) Tests of the General Linear Hypothesis

Consider the observation model described in the previous section. Instead of a confidence ellipsoid for $y = Gx$, we seek a test of the hypothesis

$$\mathscr{H}_0\colon Gx = 0.$$

In the case where σ^2 is known, it turns out that the appropriate likelihood ratio test is to reject $\mathscr{H}_0$ for large values of

(6.4.1) $$\boldsymbol{\rho}_1 = \sigma^{-2}\{\|\mathbf{z}-H\bar{\mathbf{x}}\|^2 - \|\mathbf{z}-H\hat{\mathbf{x}}\|^2\}$$

where $\hat{\mathbf{x}}$ is the least squares estimator for x:

(6.4.2) $$\hat{\mathbf{x}} = H^+\mathbf{z}$$

and $\bar{\mathbf{x}}$ is the least squares estimator for x, computed subject to the constraint that $Gx = 0$:

(6.4.3) $$\bar{\mathbf{x}} = \bar{H}^+\mathbf{z}$$

where

$$\bar{H} = H(I-G^+G). \tag{3.12.4}$$

If σ^2 is not known, the likelihood ratio test rejects $\mathscr{H}_0$ for large values of

(6.4.4) $$\boldsymbol{\rho}_2 = \|\mathbf{z}-H\bar{\mathbf{x}}\|^2/\|\mathbf{z}-H\hat{\mathbf{x}}\|^2.$$

(6.4.5) **Exercise:** If $\hat{\mathbf{x}}^*$ and $\bar{\mathbf{x}}^*$ are any other least squares estimate and constrained least squares estimate (computed subject to $Gx = 0$), show that

$$H\hat{\mathbf{x}}^* = H\hat{\mathbf{x}} \qquad \text{and} \qquad H\bar{\mathbf{x}}^* = H\bar{\mathbf{x}}$$

so that the same test statistics, $\boldsymbol{\rho}_1$ and $\boldsymbol{\rho}_2$, result. [*Hint:* Use (3.12.2) and (3.12.4).]

We will now prove that

(6.4.6) $\boldsymbol{\rho}_1$ has a noncentral chi-square distribution with $\tau\,[=\text{rk}(HH^+ - \bar{H}\bar{H}^+)]$ degrees of freedom and noncentrality parameter δ where

(6.4.7)
$$\delta = \sigma^{-1}\,\|(HH^+ - \bar{H}\bar{H}^+)Hx\|$$

and

(6.4.8)
$$\frac{n-k_2}{\tau}(\boldsymbol{\rho}_2 - 1) \equiv \frac{n-k_2}{\tau}\left[\frac{\|\mathbf{z}-H\bar{\mathbf{x}}\|^2 - \|\mathbf{z}-H\hat{\mathbf{x}}\|^2}{\|\mathbf{z}-H\hat{\mathbf{x}}\|^2}\right]$$

has a noncentral F distribution with τ and $n-k_2$ degrees of freedom $[k_2 = \text{rk}(H)]$ and noncentrality parameter δ.

Furthermore,

(6.4.9)
$$\delta = 0 \qquad \text{if} \quad Gx = 0$$

so that the distributions of $\boldsymbol{\rho}_1$ and $\boldsymbol{\rho}_2$ are central chi-square and F when the null hypothesis is true.

To prove (6.4.7), we note that

(6.4.10)
$$\mathscr{R}(\bar{H}) \subseteq \mathscr{R}(H)$$

so that $(HH^+)(\bar{H}\bar{H}^+) = \bar{H}\bar{H}^+$. After a slight amount of algebra, it follows that

(6.4.11)
$$\begin{aligned}\|(HH^+ - \bar{H}\bar{H}^+)\mathbf{z}\|^2 &= \|(\mathbf{z}-\bar{H}\bar{H}^+\mathbf{z}) - (\mathbf{z}-HH^+\mathbf{z})\|^2\\ &= \|(\mathbf{z}-\bar{H}\bar{\mathbf{x}}) - (\mathbf{z}-H\hat{\mathbf{x}})\|^2\\ &= \|\mathbf{z}-\bar{H}\bar{\mathbf{x}}\|^2 - \|\mathbf{z}-H\hat{\mathbf{x}}\|^2.\end{aligned}$$

Thus

$$\boldsymbol{\rho}_1 = \|\sigma^{-1}(HH^+ - \bar{H}\bar{H}^+)\mathbf{z}\|^2.$$

The random variable $\sigma^{-1}\mathbf{z}$ is normally distributed with mean $\sigma^{-1}Hx$ and covariance I. Since $\mathscr{R}(\bar{H}) \subseteq \mathscr{R}(H)$, $HH^+ - \bar{H}\bar{H}^+$ is the projection on $\mathscr{R}(H) - \mathscr{R}(\bar{H})$, (2.7.4b), and so (6.2.6) guarantees that

$$\|\sigma^{-1}(HH^+ - \bar{H}\bar{H}^+)\mathbf{z}\|^2$$

has a noncentral chi-square distribution with noncentrality parameter δ and τ df. This establishes (6.4.7).

To prove (6.4.8), it suffices to point out that $\mathbf{z}-H\hat{\mathbf{x}} = (I-HH^+)\mathbf{z}$ is is uncorrelated with $(HH^+ - \bar{H}\bar{H}^+)\mathbf{z}$ because

$$\begin{aligned}\mathscr{E}(\mathbf{z}-H\hat{\mathbf{x}})[(HH^+ - \bar{H}\bar{H}^+)(\mathbf{z}-Hx)]^{\mathrm{T}} &= (I-HH^+)\mathscr{E}\mathbf{v}\mathbf{v}^{\mathrm{T}}(HH^+ - \bar{H}\bar{H}^+)\\ &= 0\end{aligned}$$

because $\mathscr{E}\mathbf{v}\mathbf{v}^{\mathrm{T}} = \sigma^2 I$ and $HH^+(\bar{H}\bar{H}^+) = \bar{H}\bar{H}^+$, (6.4.10). Therefore $\|\mathbf{z}-H\hat{\mathbf{x}}\|^2$ is independent of $\|(HH^+ - \bar{H}\bar{H}^+)\mathbf{z}\|^2 = \|\mathbf{z}-\bar{H}\bar{\mathbf{x}}\|^2 - \|\mathbf{z}-H\hat{\mathbf{x}}\|^2$, (6.4.11), and we showed in (6.3) that $(\sigma^2)^{-1}\|\mathbf{z}-H\hat{\mathbf{x}}\|^2$ has a central chi-square distribution with $n-k_2$ df, where $k_2 = \mathrm{rk}(H)$. The result now follows from (6.2.3).

To prove (6.4.9), we note that $x = (I-G^+G)x$ if $Gx = 0$ so that

$$\begin{aligned}(HH^+ - \bar{H}\bar{H}^+)Hx &= HH^+Hx - \bar{H}\bar{H}^+Hx\\ &= Hx - \bar{H}\bar{H}^+H(I-G^+G)x\\ &= H(I-G^+G)x - \bar{H}\bar{H}^+\bar{H}x\\ &= 0\end{aligned}$$

which implies

$$\delta = 0.$$

If $\delta > 0$, the values of $\boldsymbol{\rho}_1$ and $\boldsymbol{\rho}_2$ tend to be larger than if $\delta = 0$. To test $\mathscr{H}_0$ at the significance level α, reject $\mathscr{H}_0$ if $\boldsymbol{\rho}_1 > \omega_{1-\alpha}$ [the upper 100 $(1-\alpha)$ percentile of the central chi-square distribution with τ df] provided σ is known. Otherwise, reject $\mathscr{H}_0$ if $\tau^{-1}(m-k_2)(\boldsymbol{\rho}_2 - 1) > \tilde{\omega}_{1-\alpha}$, the upper $100(1-\alpha)$ percentile of the central F distribution with τ and $n-k_2$ df. The power of both tests against various alternatives is a function of δ only, and may be computed using the appropriate noncentral F and chi-square tables.

(6.5) The Relationship between Confidence Ellipsoids for *Gx* and Tests of the General Linear Hypothesis

We will now show that the likelihood ratio tests which we developed in (6.4) (σ known and σ unknown) are equivalent to the test procedures which reject the hypothesis $Gx = 0$ whenever the $1-\alpha$-level confidence ellipsoid (σ known, σ unknown) for Gx fails to cover the point 0. (Recall, we developed two ellipsoids; one for known sigma, one for unknown sigma.) This result, plus an alternate representation for δ, the noncentrality parameter, and τ, the rank of $HH^+ - \bar{H}\bar{H}^+$ follow easily from

(6.5.1) **Theorem:** If $\mathscr{R}(G^{\mathrm{T}}) \subseteq \mathscr{R}(H^{\mathrm{T}})$ and $\overline{H} = H(I-G^{+}G)$ then

(a) $\operatorname{rk}[H(I-G^{+}G)] = \operatorname{rk}(H) - \operatorname{rk}(G)$

and

(b) $(GH^{+})^{+}(GH^{+}) = HH^{+} - \overline{H}\overline{H}^{+}$

Proof: (a)
$$\operatorname{rk}[H(I-G^{+}G)] = \operatorname{rk}[H^{+}H(I-G^{+}G)] \qquad (4.10.2b)$$
$$= \operatorname{rk}[H^{+}H-G^{+}G]$$

since $H^{+}HG^{+}G = G^{+}G$ whenever $\mathscr{R}(G^{\mathrm{T}}) \subseteq \mathscr{R}(H^{\mathrm{T}})$, (3.5), so that

$$\operatorname{rk}[H(I-G^{+}G)] = \operatorname{rk}(H^{+}H-G^{+}G)$$
$$= \operatorname{rk}(H^{+}H) - \operatorname{rk}(G^{+}G) \qquad (4.10.3)$$
$$= \operatorname{rk}(H) - \operatorname{rk}(G). \qquad (4.10.2)$$

(b) Let

$$P_1 = (GH^{+})^{+}(GH^{+})$$

and

$$P_2 = \overline{H}\overline{H}^{+}.$$

$$\mathscr{R}(P_1) = \mathscr{R}(H^{+\mathrm{T}}G^{\mathrm{T}}) \subseteq \mathscr{R}(H^{+\mathrm{T}}) = \mathscr{R}(H) = \mathscr{R}(HH^{+}) \quad \text{(3.5) and (3.11.5)}$$

and

$$\mathscr{R}(P_2) = \mathscr{R}(\overline{H}) \subseteq \mathscr{R}(H) = \mathscr{R}(HH^{+}).$$

By virtue of the fact that $\mathscr{R}(G^{\mathrm{T}}) \subseteq \mathscr{R}(H^{\mathrm{T}})$, it follows that

(6.5.1.1)
$$G = GH^{+}H. \qquad \text{(6.3.1) and (6.3.2)}$$

Hence

$$GH^{+}\overline{H} = GH^{+}H(I-G^{+}G) = 0.$$

Therefore $P_1 P_2 = 0$ and by (4.10.3), it suffices to show that $\operatorname{rk}(P_1)+\operatorname{rk}(P_2) = \operatorname{rk}(HH^{+})$. It will then follow immediately that $HH^{+} = P_1+P_2$.
But

$$\operatorname{rk}(P_1) = \operatorname{rk}(GH^{+}) = \operatorname{rk}(GH^{+}H) = \operatorname{rk}(G) \quad \text{(6.5.1.1) and (4.10.2)}$$

and

$$\operatorname{rk}(P_2) = \operatorname{rk}(\overline{H}) = \operatorname{rk}[H(I-G^{+}G)]$$
$$= \operatorname{rk}(H) - \operatorname{rk}(G) \qquad \text{[part (a)]}.$$

Thus,

$$\operatorname{rk}(P_1) + \operatorname{rk}(P_2) = \operatorname{rk}(H) = \operatorname{rk}(HH^{+}) \qquad (4.10.2)$$

and this proves (b), (4.10.3b). ■

(6.5.2) **Corollary:** Let δ and τ be as defined in (6.4.6) and (6.4.7). If Gx is an evpf, then $\tau = \mathrm{rk}(G)$ and $\delta = \sigma^{-1}\|(GH^+)^+Gx\|$.

Proof

$$\tau = \mathrm{rk}(HH^+ - \bar{H}\bar{H}^+).$$

If G is an evpf, then $\mathscr{R}(G^\mathrm{T}) \subseteq \mathscr{R}(H^\mathrm{T})$, (6.3.1), and by (6.5.1b) and (6.3.6.1),

$$\tau = \mathrm{rk}[(GH^+)^+(GH^+)] = \mathrm{rk}(GH^+) = \mathrm{rk}(G).$$

By the same token,

$$\begin{aligned}\delta &= \sigma^{-1}\|(HH^+ - \bar{H}\bar{H}^+)Hx\| \\ &= \sigma^{-1}\|(GH^+)^+(GH^+Hx)\| \\ &= \sigma^{-1}\|(GH^+)^+Gx\| \qquad \text{if } G \text{ is an evpf.} \;\blacksquare \qquad (6.3.2)\end{aligned}$$

Referring to section (6.3), we now see that the confidence ellipsoid $\mathscr{S}(\sigma^2\omega_{1-\alpha})$ covers the point $y = 0$ if and only if

(6.5.3) $$\hat{\mathbf{y}}^\mathrm{T}[G(H^\mathrm{T}H)^+G^\mathrm{T}]^+\hat{\mathbf{y}} \leqslant \sigma^2\omega_{1-\alpha}$$

where $\omega_{1-\alpha}$ is the upper $100(1-\alpha)$ percentile of the chi-square distribution with τ df. Since $\hat{\mathbf{y}} = GH^+\mathbf{z}$, we can write (6.5.3) as

(6.5.4) $$\|(GH^+)^+GH^+\mathbf{z}\|^2 \leqslant \sigma^2\omega_{1-\alpha}$$

which, by virtue of (6.5.1b), is the same as

(6.5.5) $$(\sigma^2)^{-1}\|(HH^+ - \bar{H}\bar{H}^+)\mathbf{z}\|^2 \leqslant \omega_{1-\alpha}.$$

Thus, the likelihood ratio test for the case of known σ, rejects $\mathscr{H}_0$ at the level α if and only if the $(1-\alpha)$ level confidence ellipsoid for Gx fails to cover the point $y = 0$.

(6.5.6) **Exercise:** Show that the likelihood ration test (for unknown σ) at the level α rejects $\mathscr{H}_0$ if and only if

$$\mathscr{S}(k_1\hat{\boldsymbol{\sigma}}^2\tilde{\omega}_{1-\alpha}) \qquad \text{[defined in (6.3)]}$$

fails to cover $y = 0$.

(6.5.7) **Exercise:** If G is an evpf then $\delta = 0$ if and only if $Gx = 0$.

(6.6) Orthogonal Designs

In statistical investigations, it is often desired to obtain information about the value of one or more parameters and in an effort to do so, the researcher designs an experiment which will yield data of the form

$$\mathbf{z} = Hx + \mathbf{v}$$

where $\mathbf{z}$ is the "observation vector," x is an unknown vector whose components include the parameters of interest (and possibly other so-called "nuisance parameters") $\mathbf{v}$ is the vector of residuals (or "observation errors"). H is a known matrix, often referred to as the "experimental design." The art of experimental design occupies an important niche in the lore of applied statistics and one of its basic concepts is the notion of "orthogonal design."

Suppose $y_j = G_j x$ are evpf's for $j = 1, 2, \ldots, p$ (i.e., each component of each y_j is an estimable parametric function).

The design H is said to be *orthogonal with respect to the evpf's* $G_1 x, G_2 x, \ldots, G_p x$ if and only if the BLUE's for y_i and y_j are uncorrelated when $i \neq j$.

One reason that orthogonal designs are desirable is to be explained in Theorem (6.6.9). The BLUE for each y_j can be computed as though all the other y_i's are known to be zero. This fact often reduces the computational complexity of a least squares problem by several orders of magnitude.

For the sake of clarity, we will confine our attention to the case where the residual vector $\mathbf{v}$ has covariance proportional to the identity matrix. In that case, the BLUE for y_j is

(6.6.1) $$\hat{\mathbf{y}}_j = G_j H^+ \mathbf{z} \qquad (j = 1, \ldots, p) \tag{6.1.12}$$

and

$$\begin{aligned}
\operatorname{Cov}(\hat{\mathbf{y}}_i, \hat{\mathbf{y}}_j) &\equiv \mathscr{E}(\hat{\mathbf{y}}_i - y_i)(\hat{\mathbf{y}}_j - y_j)^{\mathrm{T}} \\
&= \mathscr{E} G_i H^+(\mathbf{z} - Hx)(\mathbf{z} - Hx)^{\mathrm{T}}(G_j H^+)^{\mathrm{T}} \\
&= G_i H^+ \mathscr{E}\mathbf{v}\mathbf{v}^{\mathrm{T}} H^{+\mathrm{T}} G_j^{\mathrm{T}} \\
&= \sigma^2 G_i H^+ H^{+\mathrm{T}} G_j^{\mathrm{T}} \\
&= \sigma^2 G_i (H^{\mathrm{T}} H)^+ G_j^{\mathrm{T}}.
\end{aligned}$$

Thus, $\hat{\mathbf{y}}_i$ is uncorrelated with $\hat{\mathbf{y}}_j$ if and only if

(6.6.2) $$G_i (H^{\mathrm{T}} H)^+ G_j^{\mathrm{T}} = 0$$

which is the same as saying the row vectors of G_i must be "orthogonal" to the row vectors of G_j with respect to the inner product induced by $(H^{\mathrm{T}} H)^+$ (which, by the way, is proportional to the covariance of $\hat{\mathbf{x}} = H^+ \mathbf{z}$, the BLUE for x). Equation (6.6.2) can be written in the alternate form

(6.6.3) $$(G_i H^+)(G_j H^+)^{\mathrm{T}} = 0 \qquad i \neq j$$

so that H is orthogonal with respect to $G_1 x, G_2 x, \ldots, G_p x$ if and only if

$$\mathscr{R}[(G_j H^+)^{\mathrm{T}}] \subseteq \mathscr{N}(G_i H^+) \qquad \text{when} \quad i \neq j.$$

Since

$$\mathscr{R}[(G_j H^+)^{\mathrm{T}}] = \mathscr{R}[(G_j H^+)^+ (G_j H^+)]$$

and

$$\mathscr{N}(G_i H^+) = \mathscr{N}[(G_i H^+)^+ (G_i H^+)] \qquad (3.11.5)$$

we see that H is orthogonal with respect to $G_1 x, \ldots, G_p x$ if and only if

(6.6.4) $$P_i P_j = 0 \qquad \text{when} \quad i \neq j$$

where

(6.6.5) $$P_j = (G_j H^+)^+ (G_j H^+) \qquad j = 1, 2, \ldots, p.$$

In the proof of (6.5.1b) we showed that

$$\mathscr{R}[(GH^+)^+ (GH^+)] \subseteq \mathscr{R}(HH^+)$$

if Gx is an evpf, so it follows that

(6.6.6) $$\mathscr{R}\left[\sum_{j=1}^{p} (G_j H^+)^+ (G_j H^+)\right] \subseteq \mathscr{R}(HH^+).$$

If equality holds in (6.6.6), then by (3.7.7), HH^+ is expressible as the sum of *orthogonal projections*, a fact which we will see to be of more than passing interest. [Two projections P_1 and P_2 are said to be *orthogonal* if $P_1 P_2 = 0$. Clearly $P_1 P_2 = 0$ if and only if $P_2 P_1 = 0$ and the last is true if and only if $\mathscr{R}(P_1) \subseteq \mathscr{N}(P_2)$.]

If strict inclusion holds in (6.6.6) (as it may) we can, by suitably defining one more evpf (which we call $y_0 = G_0 x$), decompose HH^+ into a sum of orthogonal projections:

$$HH^+ = \sum_{j=0}^{p} (G_j H^+)^+ (G_j H^+).$$

In the next theorem, we show how to define G_0:

(6.6.7) **Theorem:** If H is an orthogonal design with respect to $G_1 x, \ldots, G_p x$ and if

(6.6.7.1) $$T = HH^+ - \sum_{j=1}^{p} (G_j H^+)^+ (G_j H^+) \neq 0$$

then H is also an orthogonal design with respect to $G_0 x, G_1 x, \ldots, G_p x$ where

$$G_0 = TH.$$

Furthermore,

$$HH^+ = \sum_{j=0}^{p} (G_j H^+)^+ (G_j H^+) \tag{6.6.7.2}$$

and the projections in the sum are mutually orthogonal.

Proof

$$G_0 H^+ H = THH^+ H = TH = G_0$$

so G_0 is an evpf, (6.3.2).

To show that H is orthogonal with respect to $G_0 x, G_1 x, \ldots, G_p x$, we must show that

$$(G_k H^+)(G_0 H^+)^{\mathrm{T}} = 0 \qquad k = 1, \ldots, p.$$

The left side is

$$(G_k H^+)(THH^+)^{\mathrm{T}} = G_k H^+ T$$

(since $H^+HH^+ = H^+$ and T is symmetric). If $G_k H^+$ is applied to each term on the right side of (6.6.7.1), we see that

$$\begin{aligned} G_k H^+ T &= G_k H^+ [HH^+ - (G_k H^+)^+ (G_k H^+)] \\ &= G_k H^+ - G_k H^+ = 0. \end{aligned}$$

This shows that H is orthogonal with respect to $G_0 x, \ldots, G_p x$. Since $G_0 = TH$,

$$[G_0 H^+]^{\mathrm{T}} = [THH^+]^{\mathrm{T}} = HH^+ T$$

and since $\mathscr{R}(T) \subseteq \mathscr{R}(HH^+)$, (6.6.6),

$$HH^+ T = T. \tag{3.13.1}$$

Thus,

$$T = T^{\mathrm{T}} = (G_0 H^+)^{\mathrm{T}}$$

and since T is the projection on $\mathscr{R}(HH^+) - \mathscr{R}(P_1 + P_2 + \cdots + P_p)$,

$$(G_0 H^+)^+ = T^+ = T$$

so that

$$T = T^2 = (G_0 H^+)^+ (G_0 H^+)$$

which is another way of expressing (6.6.7.2). ■

Comment: The zeroth term of the sum (6.6.7.2) has an interpretation: Recall that the set of all estimable parametric functions consists of all parameters of the form $g^{\mathrm{T}} x$ where g is free to range over $\mathscr{R}(H^+H)$. The components of the evpf's $G_1 x, G_2 x, \ldots, G_p x$ collectively correspond to the class of estimable parametric functions of the form $g^T x$ where g is free to range over $\mathscr{R}[(G_1^T \mid G_2^T \mid \cdots \mid G_p^T)]$. Since $H^+ H G_j^T = G_j^T$, this manifold is a subspace of $\mathscr{R}(H^+ H)$. Now, G_0 is defined in such a way that

$$\mathscr{R}[(G_0^T \mid G_1^T \mid \cdots \mid G_p^T)] = \mathscr{R}(H^+ H)$$

and so that H remains an orthogonal design.

(6.6.8) **Exercise:** Show that $\mathscr{R}[(G_0^T \mid G_1^T \mid \cdots \mid G_p^T)] = \mathscr{R}[H^+ H]$.

The concept of "orthogonal design" was of great importance in the days before the widespread availability of digital computers. The task of solving the least squares "normal equations" and evaluating the residual sum of squares on a desk calculator, was tedious at best and overwhelming if the dimensionality of the unknown vector x was greater than 6 or 7 *unless* the design was orthogonal. In that case, as we will see in the next theorem, each evpf could be estimated separately as though the rest were known to be zero.

The general effect was to reduce the estimation of a k-dimensional parameter (with associated computations equivalent in complexity to that of inverting a $k \times k$ matrix) to the estimation of $p+1$ parameters, each of dimension ρ_i (each requiring the equivalent of a $\rho_i \times \rho_i$ matrix inversion) where $\sum_i \rho_i = k$. Generally speaking, the inversion of a $k \times k$ matrix requires on the order of k^3 arithmetic operations. Therefore

$$k^3 = \left(\sum_i \rho_i\right)^3 >> \sum_i \rho_i^3$$

so that considerable computational savings would be realized were one to capitalize on the orthogonality of the design.

Finally, an intuitively informative decomposition for the residual sum of squares, $\|\mathbf{z} - Hx\|^2$ is another dividend associated with orthogonal designs. To be precise:

(6.6.9) **Theorem** (*Orthogonal decomposition for analysis of variance*)

(a) If $\mathbf{z} = Hx + \mathbf{v}$ where $\mathbf{v}$ is a zero-mean vector random variable with covariance $\sigma^2 I$, if $y_i = G_i x$ $(i = 1, \ldots, p)$ are evpf's, and if H is an orthogonal design with respect to these evpf's, then

$$\text{(6.6.9.1)} \quad \|\mathbf{z} - Hx\|^2 = \|\mathbf{z} - H\hat{\mathbf{x}}\|^2 + \sum_{j=0}^{p} (\hat{\mathbf{y}}_j - y_j)^T [G_j (H^T H)^+ G_j^T]^+ (\hat{\mathbf{y}}_j - y_j)$$

where G_0 is the (possibly zero) matrix defined in (6.6.7), $\hat{\mathbf{x}} = H^+ \mathbf{z}$ is the least squares estimate for x and $\hat{\mathbf{y}}_j = G_j \hat{\mathbf{x}}$ is the BLUE for y_j $(j = 0, 1, \ldots, p)$.

(b) $\hat{\mathbf{y}}_j$ can be computed in the following (alternative) fashion: Let $\hat{\mathbf{x}}_j$ be any vector minimizing $\|\mathbf{z}-Hx\|^2$ subject to the constraints

$$G_i x = 0 \qquad \text{for} \quad i = 1,2,\ldots,j-1,j+1,\ldots,p.$$

Then

$$\hat{\mathbf{y}}_j = G_j \hat{\mathbf{x}}_j. \tag{6.6.9.2}$$

(c) If it is assumed, in addition, that the residual vector $\mathbf{v}$ is normally distributed, then the first term on the right side of (6.6.9.1) (when divided by σ^2) has a chi-square distribution with $n-k$ df, where n is the dimension of $\mathbf{z}$ and k is the rank of H. The jth term of the sum (when divided by σ^2) has a chi-square distribution with ρ_j df, where ρ_j is the rank of G_j. All $p+2$ terms are mutually independent and $\sum_{i=0}^{p} \rho_i = k$.

Comment: The sum of squares decomposition (6.6.9.1), "explains" the "total" sum of squares $\|\mathbf{z}-Hx\|^2$ in terms of the sum of squares associated with each of the evpf's y_j and the so-called residual sum of squares $\|\mathbf{z}-H\hat{\mathbf{x}}\|^2$. These quantities are traditionally computed with the y_j's set equal to zero and they are exhibited under the title "analysis of variance table" (Scheffè [1, Chap. 4]).

Proof: (a) Since

$$\begin{aligned} \mathbf{z} - Hx &= \mathbf{z} - HH^+\mathbf{z} + HH^+(\mathbf{z}-Hx) \\ &= (I-HH^+)\mathbf{z} + HH^+(\mathbf{z}-Hx) \end{aligned} \tag{6.6.9.3}$$

we deduce from (6.6.7.2) that

$$\mathbf{z} - Hx = (I-HH^+)\mathbf{z} + \sum_{j=0}^{p} (G_j H^+)^+ (G_j H^+)(\mathbf{z}-Hx). \tag{6.6.9.4}$$

The G_j's are associated with evpf's, hence

$$G_j H^+ Hx = G_j x = y_j \tag{6.3.2}$$

and since $H^+\mathbf{z} = \hat{\mathbf{x}}$ is the BLUE for H^+Hx and $G_j\hat{\mathbf{x}} = \hat{\mathbf{y}}_j$ is the BLUE for y_j, we see that

$$\mathbf{z} - Hx = (\mathbf{z}-H\hat{\mathbf{x}}) + \sum_{j=0}^{p} (G_j H^+)^+ (\hat{\mathbf{y}}_j - y_j). \tag{6.6.9.5}$$

The first term on the right side of (6.6.9.4) is uncorrelated with and orthogonal to each term in the sum since (a) $I-HH^+$ projects onto $\mathscr{R}(H)^\perp$ $[=\mathscr{N}(H^{\mathrm{T}})]$, whereas each term of the sum projects onto a subspace of $\mathscr{R}(H)$, (6.6.6), and (b) the covariance of $\mathbf{z}-Hx$ is proportional to the identity. Since H is an orthogonal design with respect to the $G_j x$'s, the projections in the sum are mutually orthogonal, (6.6.4). So then are each of the *vectors* in the sum, and since the covariance of $\mathbf{z}-Hx$ is proportional to the identity, the terms are also mutually uncorrelated.

The mutual orthogonality of all terms allows for the sum of squares decomposition for (6.6.9.5):

$$\begin{aligned}\|\mathbf{z}-Hx\|^2 &= \|\mathbf{z}-H\hat{\mathbf{x}}\|^2 + \sum_{j=0}^{p} \|(G_j H^+)^+ (\hat{\mathbf{y}}_j - y_j)\|^2 \\ &= \|\mathbf{z}-H\hat{\mathbf{x}}\|^2 + \sum_{j=0}^{p} (\hat{\mathbf{y}}_j - y_j)^{\mathrm{T}} [G_j (H^{\mathrm{T}} H)^+ G_j^{\mathrm{T}}]^+ (\hat{\mathbf{y}}_j - y_j),\end{aligned}$$

which establishes (a).

(b) If $\hat{\mathbf{x}}_j$ minimizes $\|\mathbf{z}-Hx\|^2$ subject to $G_i x = 0$ for all $i \neq j$, then

$$\hat{\mathbf{m}}_j = H\hat{\mathbf{x}}_j$$

minimizes $\|\mathbf{z}-m\|^2$ subject to $G_i H^+ m = 0$ for all $i \neq j$ and $m \in \mathscr{R}(H)$, since $G_i H^+ m = G_i H^+ Hx$ and $G_i H^+ Hx = G_i x$ when G_i is an evpf, (6.3.2).

Therefore, $\hat{\mathbf{m}}_j$ minimizes $\|\mathbf{z}-m\|^2$ subject to the constraint

(6.6.9.6) $$(I - HH^+)m + \sum_{\substack{i=0 \\ i \neq j}}^{p} (G_i H^+)^+ (G_i H^+) m = 0,$$

since

$$G_i H^+ m = 0$$

if and only if

$$(G_i H^+)^+ (G_i H^+) m = 0$$

and since the terms in (6.6.9.6) are mutually orthogonal.

By (6.6.7.2),

(6.6.9.7) $$(I - HH^+) + \sum_{\substack{i=0 \\ i \neq j}}^{p} (G_i H^+)^+ (G_i H^+) = I - (G_j H^+)^+ (G_j H^+) \equiv Q_j$$

which is a projection, so that

$$\|\mathbf{z}-m\|^2 = \|\mathbf{z} - Q_j m - (I - Q_j) m\|^2$$

is minimized subject to the constraint $Q_j m = 0$ if and only if

(6.6.9.8) $$\hat{\mathbf{m}}_j = (I - Q_j)\mathbf{z}.$$ (3.12.4)

Since $\hat{\mathbf{m}}_j = H\hat{\mathbf{x}}_j$, we see that

$$G_j H^+ \hat{\mathbf{m}}_j = G_j H^+ H\hat{\mathbf{x}}_j = G_j \hat{\mathbf{x}}_j$$ (6.3.2)

whereas (6.6.9.8) implies that

$$G_j H^+ \hat{\mathbf{m}}_j = G_j H^+ [(G_j H^+)^+ (G_j H^+)] \mathbf{z}$$
$$= G_j H^+ \mathbf{z} = G_j \hat{\mathbf{x}}.$$

Hence

$$G_j \hat{\mathbf{x}} = G_j \hat{\mathbf{x}}_j = \hat{\mathbf{y}}_j$$

which proves (6.6.9.2).

(c) In part (a), we showed that all terms on the right side of (6.6.9.5) are uncorrelated. If a normality assumption is added, they are independent as well, (6.2.4), which implies that all terms on the right side of (6.6.9.1) are mutually independent. The distribution of $\sigma^{-2}\|\mathbf{z}-H\hat{\mathbf{x}}\|^2$ has been shown to be chi-square with $n-k$ df [(6.3.10), (6.3.11)]. The distribution of $\sigma^{-2}(\hat{\mathbf{y}}_j-y_j)^{\mathrm{T}}[G_j(H^{\mathrm{T}}H)^+G_j^{\mathrm{T}}](\hat{\mathbf{y}}_j-y_j)$ was shown to be chi-square with ρ_j df in (6.3.5).

The left side of (6.6.9.1), when divided by σ^2, is of the form

$$\sum_{j=1}^{n} (\mathbf{v}_j/\sigma)^2$$

where the $\mathbf{v}_j$'s are independent with mean zero and variance σ^2, so that the left side of (6.6.9.1), when divided by σ^2 is chi-square with n df. Thus,

$$n = n - k + \sum_{j=0}^{p} \rho_j$$

so that

$$\sum_{j=0}^{p} \rho_j = k.$$

Comment: By definition, H is always orthogonal with respect to a single evpf $G_1 x$. This means that a decomposition of the form (6.6.9.1) can be exhibited for the case $p = 1$, provided G_0 is properly defined [according to (6.6.7)].

(6.6.10) **Exercise:** (a) H is orthogonal with respect to $G_i x$ $(i = 1, 2, \ldots, p)$ if and only if

$$(HH^+ - \bar{H}_i \bar{H}_i^+)(HH^+ - \bar{H}_j \bar{H}_j^+) = 0$$

when $i \neq j$, where $\bar{H}_j = H(I - G_j G_j^+)$. [*Hint:* Use (6.5.1b).]

(b) Let x_j^* be any x which minimizes $\|z - Hx\|^2$ subject to

$$x \in \mathcal{N}(G_1) \cap \cdots \cap \mathcal{N}(G_{j-1}) \cap \mathcal{N}(G_{j+1}) \cap \cdots \cap \mathcal{N}(G_p).$$

Let $y_j^* = G_j x_j^*$ and let $\hat{y}_j = G_j \hat{x}$, the BLUE for $y_j = G_j x$. H is orthogonal with respect to $G_i x$ $(i = 1, 2, \ldots, p)$ if and only if

$$y_j^* = \hat{y}_j \qquad \text{for} \quad j = 1, 2, \ldots, p.$$

(6.6.11) **Example** (*Straight line regression*): Observations are taken, of the form

$$\boldsymbol{\zeta}_i = \alpha + \beta\tau_i + \mathbf{v}_i \qquad i = 1, \ldots, n.$$

In the vector-matrix notation,

$$\mathbf{z} = Hx + \mathbf{v}$$

where

$$\mathbf{z} = \begin{pmatrix} \zeta_1 \\ \vdots \\ \zeta_n \end{pmatrix} \qquad H = (e \,\vdots\, t) \qquad x = \begin{pmatrix} \alpha \\ \beta \end{pmatrix}$$

$$e = \begin{pmatrix} 1 \\ \vdots \\ 1 \end{pmatrix} \qquad t = \begin{pmatrix} \tau_1 \\ \vdots \\ \tau_n \end{pmatrix} \qquad \text{and} \qquad \mathbf{v} = \begin{pmatrix} \mathbf{v}_1 \\ \vdots \\ \mathbf{v}_n \end{pmatrix}.$$

If we let $G_1 = (1 \quad 0)$ and $G_2 = (0 \quad 1)$, then $y_1 = G_1 x = \alpha$ and $y_2 = G_2 x = \beta$. H has rank $k = 2$ if the τ's are not all the same, in which case $H^+ H = I$ so y_1 and y_2 are evpf's.

$$\bar{H}_j \equiv H(I - G_j G_j^+) = \begin{cases} (0 \,\vdots\, t) & j = 1 \\ (e \,\vdots\, 0) & j = 2 \end{cases}$$

so that

$$\bar{H}_j \bar{H}_j^+ = \begin{cases} tt^{\mathrm{T}}/\|t\|^2 & j = 1 \\ ee^{\mathrm{T}}/\|e\|^2 & j = 2. \end{cases}$$

The quickest way to find HH^+ is to perform a Gramm–Schmidt orthogonalization on the columns of H and then apply (3.7.3):

$$HH^+ = \frac{ee^{\mathrm{T}}}{\|e\|^2} + \frac{uu^{\mathrm{T}}}{\|u\|^2} \qquad \text{where} \quad u = t - \frac{(e^{\mathrm{T}} t)}{\|e\|^2} e$$

$$= \frac{tt^{\mathrm{T}}}{\|t\|^2} + \frac{ww^{\mathrm{T}}}{\|w\|^2} \qquad \text{where} \quad w = e - \frac{(t^{\mathrm{T}} e)}{\|t\|^2} t.$$

Thus

$$HH^+ - \bar{H}_1 \bar{H}_1^+ = ww^{\mathrm{T}}/\|w\|^2$$

$$HH^+ - \bar{H}_2 \bar{H}_2^+ = uu^{\mathrm{T}}/\|u\|^2$$

and so

$$(HH^+ - \bar{H}_1 \bar{H}_1{}^+)(HH^+ - \bar{H}_2 \bar{H}_2{}^+) = \frac{w^T u}{\|u\|^2 \|w\|^2}(wu^T).$$

By virtue of (6.6.10) H is an orthogonal design with respect to $G_1 x$, $G_2 x$ if and only if the last expression vanishes. Since

$$w^T u = (e^T t)\left[\frac{(e^T t)^2 - \|e\|^2 \|t\|^2}{\|e\|^2 \|t\|^2}\right]$$

and since t is not a multiple of e if the τ's are not all the same, Schwarz's inequality tells us that the term in square brackets is strictly negative so $w^T u = 0$ if and only if $(e^T t) = 0$. In summary, H is orthogonal with respect to α and β if and only if $n^{-1}\sum_{j=1}^n \tau_j \equiv \bar{\tau} = 0$. So, H is an orthogonal design if and only if the observations are chosen at values of τ_j whose average value is zero. In that case, we can compute the BLUE for α as though $\beta = 0$:

"Pretend" that

$$\zeta_i = \alpha + \mathbf{v}_i \qquad i = 1, \ldots, n.$$

Then the BLUE for α is

$$\hat{\boldsymbol{\alpha}} = n^{-1} \sum_{i=1}^n \zeta_i.$$

Similarly, we can compute the BLUE for β as though $\alpha = 0$:

"Pretend" that

$$\zeta_i = \beta\tau_i + \mathbf{v}_i \qquad i = 1, \ldots, n.$$

The BLUE for β is

$$\hat{\boldsymbol{\beta}} = t^+ \mathbf{z} = \sum_{j=1}^n \zeta_j \tau_j \Big/ \sum_{j=1}^n \tau_j^2.$$

In this special two-dimensional problem, the BLUE for x can be exhibited explicitly in the general (nonorthogonal) case:

$$\hat{\mathbf{x}} = \begin{pmatrix} \hat{\boldsymbol{\alpha}} \\ \hat{\boldsymbol{\beta}} \end{pmatrix}$$

where

$$\hat{\boldsymbol{\beta}} = \sum_{j=1}^n (\zeta_j - \bar{\zeta})(\tau_j - \bar{\tau}) \Big/ \sum_{j=1}^n (\tau_j - \bar{\tau})^2$$

$$\bar{\boldsymbol{\zeta}} = n^{-1} \sum_{j=1}^n \zeta_j \qquad \bar{\tau} = n^{-1} \sum_{j=1}^n \tau_j$$

and

$$\hat{\boldsymbol{\alpha}} = \bar{\boldsymbol{\zeta}} - \hat{\boldsymbol{\beta}}\bar{\tau}.$$

It is apparent that the orthogonality condition $\bar{\tau} = 0$, reduces these formulas to the estimates for α and β given previously.

The sum of squares decomposition for the *orthogonal case* is

$$\sigma^{-2} \sum_{j=1}^{n} (\zeta_j - \alpha - \beta\tau_j)^2 = \sigma^{-2} \sum_{j=1}^{n} (\zeta_j - \hat{\boldsymbol{\alpha}} - \hat{\boldsymbol{\beta}}\tau_j)^2 + (\hat{\boldsymbol{\beta}} - \beta)^2/\sigma_{\beta}{}^2 + (\hat{\boldsymbol{\alpha}} - \alpha)^2/\sigma_{\hat{\alpha}}{}^2$$

where

$$\hat{\boldsymbol{\alpha}} = n^{-1} \sum_{i=1}^{n} \zeta_i \qquad \sigma_{\hat{\alpha}}{}^2 = \sigma^2/n$$

$$\hat{\boldsymbol{\beta}} = \sum_{i=1}^{n} \zeta_i \tau_i \Big/ \sum_{i=1}^{n} \tau_i{}^2 \qquad \sigma_{\beta}{}^2 = \sigma^2 \Big/ \sum_{i=1}^{n} \tau_i{}^2.$$

If the design were not orthogonal with respect to α and β, it would still be orthogonal to either one, taken separately. Suppose β were the parameter of interest. Let us redefine

$$G_1 = (0 \quad 1)$$

so that

$$G_1 x = \beta.$$

Using (6.6.7), we define

$$G_0 = [HH^+ - (G_1 H^+)^+ (G_1 H^+)] H = e(n \mid n\bar{\tau}).$$

Thus,

$$G_0 x = n(\alpha + \bar{\tau}\beta) e$$

and H is orthogonal with respect to $G_0 x$, $G_1 x$. We can compute $\hat{\boldsymbol{\beta}}$ by minimizing $\|\mathbf{z} - Hx\|^2$ subject to $G_0 x = 0$ (i.e., $\alpha = -\bar{\tau}\beta$). In this case, the sum of squares

$$\sum_{j=1}^{n} (\zeta_j - \alpha - \beta\tau j)^2$$

becomes

$$\sum_{j=1}^{n} (\zeta_j + \bar{\tau}\beta - \tau_j \beta)^2 = \sum_{j=1}^{n} [\zeta_j - \beta(\tau_j - \bar{\tau})]^2$$

which is minimized when

$$\hat{\boldsymbol{\beta}} = \sum_{j=1}^{n} \zeta_j(\tau_j - \bar{\tau}) \Big/ \sum_{j=1}^{n} (\tau_j - \bar{\tau})^2$$

$$= \sum_{j=1}^{n} (\zeta_j - \bar{\boldsymbol{\zeta}})(\tau_j - \bar{\tau}) \Big/ \sum_{j=1}^{n} (\tau_j - \bar{\tau})^2$$

as it should.

(6.6.12) **Exercise** (*Two-way layout*): Suppose observations are made of the form

$$\zeta_{ij} = \alpha_i + \beta_j + \mathbf{v}_{ij} \qquad i = 1,\ldots,I \quad j = 1,\ldots,J$$

where the α's and β's are not known and the residuals, $\mathbf{v}_{ij}$, have zero mean and are uncorrelated with common variance, σ^2.

$$z = \begin{bmatrix} \zeta_{11} \\ \zeta_{12} \\ \vdots \\ \zeta_{1J} \\ \zeta_{21} \\ \vdots \\ \zeta_{2J} \\ \vdots \\ \zeta_{I1} \\ \vdots \\ \zeta_{IJ} \end{bmatrix} \qquad x = \begin{bmatrix} \alpha_1 \\ \vdots \\ \alpha_I \\ \beta_1 \\ \vdots \\ \beta_J \end{bmatrix} \qquad \mathbf{v} = \begin{bmatrix} \mathbf{v}_{11} \\ \mathbf{v}_{12} \\ \vdots \\ \mathbf{v}_{1J} \\ \mathbf{v}_{21} \\ \vdots \\ \mathbf{v}_{2J} \\ \vdots \\ \mathbf{v}_{I1} \\ \vdots \\ \mathbf{v}_{IJ} \end{bmatrix}$$

and

$$H = \begin{array}{c} \\ J\left\{\begin{array}{c}\\ \\ \\ \end{array}\right. \\ \\ \\ \\ \\ J\left\{\begin{array}{c}\\ \\ \\ \end{array}\right. \end{array} \begin{bmatrix} \overbrace{1 \;\; 0 \cdots 0}^{I} & \overbrace{1 \;\; 0 \cdots 0}^{J} \\ 1 \;\; 0 \cdots 0 & 0 \;\; 1 \cdots 0 \\ \cdots\cdots & \cdots\cdots \\ 1 \;\; 0 \cdots 0 & 0 \;\; 0 \cdots 1 \\ \hline 0 \;\; 1 \cdots 0 & 1 \;\; 0 \cdots 0 \\ 0 \;\; 1 \cdots 0 & 0 \;\; 1 \cdots 0 \\ \sim\sim\sim\sim & \sim\sim\sim\sim \\ 0 \;\; 1 \cdots 0 & 0 \;\; 0 \cdots 1 \\ \hline 0 \;\; 0 \cdots 1 & 1 \;\; 0 \cdots 0 \\ 0 \;\; 0 \cdots 1 & 0 \;\; 1 \cdots 0 \\ \cdots\cdots & \cdots\cdots \\ 0 \;\; 0 \cdots 1 & 0 \;\; 0 \cdots 1 \end{bmatrix} \updownarrow \; I \text{ blocks of } J \text{ each}.$$

Let

$$G_1 x = \begin{bmatrix} \alpha_1 \\ \vdots \\ \alpha_I \end{bmatrix} \quad \text{and} \quad G_2 x = \begin{bmatrix} \beta_1 \\ \vdots \\ \beta_J \end{bmatrix}.$$

Show that the design is not orthogonal with respect to $G_1 x$ and $G_2 x$. (*Hint:* Show that the l.s.e. for $G_1 x$ is correlated with the l.s.e. for $G_2 x$.)

(6.6.13) (*Continued*): Let

$$\varepsilon = I^{-1} \sum_{i=1}^{I} \alpha_i + J^{-1} \sum_{j=1}^{J} \beta_j$$

$$\delta_i = \alpha_i - I^{-1} \sum_{j=1}^{I} \alpha_j \qquad i = 1, \ldots, I$$

$$\gamma_j = \beta_j - J^{-1} \sum_{k=1}^{J} \beta_k \qquad j = 1, \ldots, J$$

$$x = \begin{bmatrix} \varepsilon \\ \delta_1 \\ \vdots \\ \delta_{I-1} \\ \gamma_1 \\ \vdots \\ \gamma_{J-1} \end{bmatrix}$$

and let $\mathbf{z}$ and $\mathbf{v}$ be as in (6.6.12).

(a) Find H so that

$$\mathbf{z} = Hx + \mathbf{v}$$

(*Hint:* $\zeta_{ij} = \varepsilon + \delta_i + \gamma_j + \mathbf{v}_{ij}$, $i = 1, \ldots, I$, $j = 1, \ldots, J$.)

(b) Let

$$G_1 x = \varepsilon \qquad G_2 x = \begin{bmatrix} \delta_1 \\ \vdots \\ \delta_{I-1} \end{bmatrix} \quad \text{and} \quad G_3 x = \begin{bmatrix} \gamma_1 \\ \vdots \\ \gamma_{J-1} \end{bmatrix}.$$

Show that H is orthogonal with respect to $G_1 x$, $G_2 x$, and $G_3 x$.

(c) Define $\hat{\boldsymbol{\delta}}_I = -\sum_{j=1}^{I-1} \hat{\boldsymbol{\delta}}_j$, $\hat{\gamma}_J = -\sum_{j=1}^{J-1} \hat{\gamma}_j$. Then the orthogonal decomposition for the sum of squares for an orthogonal two-way layout is

$$\sum_{ij} (\zeta_{ij} - \varepsilon - \delta_i - \gamma_j)^2 = \sum_{ij} (\zeta_{ij} - \hat{\varepsilon} - \hat{\boldsymbol{\delta}}_i - \hat{\gamma}_j)^2$$
$$+ J \sum_i (\hat{\boldsymbol{\delta}}_i - \delta_i)^2 + I \sum_j (\hat{\gamma}_j - \gamma_j)^2$$
$$+ IJ(\hat{\varepsilon} - \varepsilon)^2.$$

If the residuals $\mathbf{v}_{ij}$ are independent and normally distributed with mean 0 and variance 1, the terms on the right have independent chi-square distributions with $IJ - (I + J - 1)$, $I - 1$, $J - 1$ and 1 df, respectively.

(6.6.14) **Exercise:** Let

$$H = {}_n(\overset{p}{H_1} \mid \overset{q}{H_2})$$

be an $n\times(p+q)$ matrix of rank $p+q$, and suppose

$$\mathbf{z} = Hx + \mathbf{v}$$

where the residual $\mathbf{v}$ has covariance I.

The least squares estimate (=BLUE) for x is unique and can be obtained via the pseudoinversion of the $n\times(p+q)$ matrix H:

$$\hat{\mathbf{x}} = H^+\mathbf{z}.$$

An alternative method which may involve less work (and which has applications to stepwise regression in Chapter VIII) goes like this:

Let

$$x = \begin{matrix} p \\ q \end{matrix}\begin{pmatrix} x_1 \\ \hdashline x_2 \end{pmatrix} \qquad G_0 = H_1 H_1{}^+ H$$

and

$$G = {}_q(\overset{p}{0} \mid \overset{q}{I}).$$

Then

(a) H is an orthogonal design with respect to $G_0 x$ and Gx.

(b) The BLUE for x_2 is unique and is given by

$$\hat{\mathbf{x}}_2 = G\mathbf{x}^*$$

where $\mathbf{x}^*$ minimizes $\|\mathbf{z}-Hx\|^2$ subject to $G_0 x = 0$.

(c) $\hat{\mathbf{x}}_2 = (Q_1 H_2)^+\mathbf{z} \equiv (H_2{}^{\mathrm{T}}Q_1 H_2)^+ H_2{}^{\mathrm{T}}Q_1\mathbf{z}$, where

$$Q_1 = I - H_1 H_1{}^+ \equiv I - H_1(H_1{}^{\mathrm{T}}H_1)^{-1}H_1{}^{\mathrm{T}}.$$

(d) The BLUE for x_1 is unique and is given by

$$\hat{\mathbf{x}}_1 = H_1{}^+(\mathbf{z}-H_2\hat{\mathbf{x}}_2) \equiv (H_1{}^{\mathrm{T}}H_1)^{-1}H_1{}^{\mathrm{T}}(\mathbf{z}-H_2\hat{\mathbf{x}}_2).$$

[*Comment:* $\hat{\mathbf{x}}_1$ and $\hat{\mathbf{x}}_2$ and hence,

$$\hat{\mathbf{x}} = \begin{pmatrix} \hat{\mathbf{x}}_1 \\ \hdashline \hat{\mathbf{x}}_2 \end{pmatrix}$$

can be obtained by way of a $q\times q$ inversion and a $p\times p$ inversion. If $p \approx q$ and both are large, a sizable computational saving can result.]

(e) Let $\mathbf{x}_1{}^* = H_1{}^+\mathbf{z}$.

Then

$$\|\mathbf{z}-H\hat{\mathbf{x}}\|^2 = \|\mathbf{z}-H_1\mathbf{x}_1{}^*\|^2 - \hat{\mathbf{x}}_2{}^{\mathrm{T}}(H_2{}^{\mathrm{T}}Q_1 H_2)\hat{\mathbf{x}}_2.$$

(Thus, the residual sum of squares associated with the BLUE for x under the "full" model

$$\mathbf{z} = H_1 x_1 + H_2 x_2 + \mathbf{v}$$

differs from the residual sum of squares associated with the BLUE for x_1 in the restricted model

$$\mathbf{z} = H_1 x_1 + \mathbf{v}$$

by the "correction factor"

$$\hat{\mathbf{x}}_2 H_2{}^{\mathrm{T}}Q_1 H_2\hat{\mathbf{x}}_2 \equiv \|Q_1 H_2\hat{\mathbf{x}}_2\|^2.)$$

Chapter VII

CONSTRAINED LEAST SQUARES, PENALTY FUNCTIONS, AND BLUE'S

(7.1) Penalty Functions

In many applications, we have seen that it is necessary to compute a weighted least squares estimator subject to linear equality constraints. That is, it is necessary to find a value of x which minimizes

(7.1.1) $$(z-Hx)^{\mathrm{T}}V^{-2}(z-Hx)$$

subject to the constraints

(7.1.2) $$Gx = u$$

where V is a known positive-definite matrix, z and u are given vectors, and H and G are given rectangular matrices.

A very general result in the theory of minimization which is associated with the "penalty function method," asserts that the value of x which minimizes

(7.1.3) $$h(x)+\lambda^{-2}g^2(x)$$

[call it $x(\lambda)$], converges (as $\lambda \to 0$) to

(7.1.4) $$x^0 = \lim_{\lambda\to 0} x(\lambda)$$

if certain mild continuity restrictions are met, and that x^0 minimizes $h(x)$ subject to the constraint $g(x) = 0$ (Butler and Martin [1]). Furthermore,

$$\lim_{\lambda \to 0} [h(x(\lambda)) + \lambda^{-2} g^2(x(\lambda))] = h(x^0) \tag{7.1.5}$$

so that the minimal value of (7.1.3) converges to the minimal value of h on the constraint set.

The term $(\lambda^{-2}) g^2(x)$ is called a "penalty function" because the minimization of $h(x) + \lambda^{-2} g^2(x)$ suffers when x lies outside the constraint set $[g(x) = 0]$ if λ is small.

In the case at hand, if we let

$$h(x) = (z - Hx)^{\mathrm{T}} V^{-2} (z - Hx) \tag{7.1.6}$$

and

$$g^2(x) = (u - Gx)^{\mathrm{T}} (u - Gx) \tag{7.1.7}$$

and if $x(\lambda)$ is a value of x which minimizes

$$(z - Hx)^{\mathrm{T}} V^{-2} (z - Hx) + \lambda^{-2} (u - Gx)^{\mathrm{T}} (u - Gx) \tag{7.1.8}$$

then it is reasonable to expect that

$$x^0 = \lim_{\lambda \to 0} x(\lambda) \tag{7.1.9}$$

exists and minimizes (7.1.6) subject to $g^2(x) = 0$ (i.e., $Gx = u$). Instead of invoking the general theorem, we will produce a self-contained proof based upon the material developed so far:

(7.1.10) **Theorem:** Let H be an $n \times m$ matrix, G be a $k \times m$ matrix and V be an $n \times n$ positive-definite matrix. Let

$$\tilde{H} = \begin{pmatrix} H \\ \hline G \end{pmatrix} \qquad \tilde{V}(\lambda) = \left(\begin{array}{c|c} V & 0 \\ \hline 0 & \lambda I \end{array}\right) \qquad [\text{order } (n+k) \times (n+k)]$$

and

$$\tilde{z} = \begin{pmatrix} z \\ \hline u \end{pmatrix}$$

where z is a given n-vector and u is a given k-vector.

Let

$$\tilde{x}(\lambda) = [\tilde{V}^{-1}(\lambda) \tilde{H}]^{+} \tilde{V}^{-1}(\lambda) \tilde{z}.$$

Then

(a) $\tilde{x}(\lambda)$ is the vector of minimum norm among those which minimize

$$(\tilde{z} - \tilde{H}x)^{\mathrm{T}} \tilde{V}^{-2}(\lambda)(\tilde{z} - \tilde{H}x) \equiv (z - Hx)^{\mathrm{T}} V^{-2} (z - Hx) + \lambda^{-2} (u - Gx)^{\mathrm{T}} (u - Gx).$$

(b) $\lim_{\lambda\to 0}\tilde{x}(\lambda) = x^0$ always exists.

(c) Among all vectors which minimize $\|u-Gx\|^2$, x^0 is the one of minimum norm among those which minimize $(z-Hx)^{\mathrm{T}}V^{-2}(z-Hx)$.

(d) If $u\in\mathscr{R}(G)$, then the set of x's which minimize $\|u-Gx\|^2$ is identical with the set of x's which satisfy the constraint $Gx = u$. In this case, x^0 minimizes $(z-Hx)^{\mathrm{T}}V^{-2}(z-Hx)$ subject to the constraint $Gx = u$.

Furthermore,

$$\lim_{\lambda\to 0}[\tilde{z}-\tilde{H}\tilde{x}(\lambda)]^{\mathrm{T}}V^{-2}(\lambda)[\tilde{z}-\tilde{H}\tilde{x}(\lambda)] = (z-Hx^0)^{\mathrm{T}}V^{-2}(z-Hx^0).$$

Proof of theorem: (a) $(\tilde{z}-\tilde{H}x)^{\mathrm{T}}\tilde{V}^{-2}(\lambda)(\tilde{z}-\tilde{H}x) = \|\tilde{V}^{-1}\tilde{z}-\tilde{V}^{-1}\tilde{H}x\|^2$ and part (a) follows from (3.4).

(b) and (c): Let

$$F = V^{-1}H \tag{7.1.10.1}$$

$$w = V^{-1}z. \tag{7.1.10.2}$$

Then by (3.8.1)

$$\begin{aligned}\tilde{x}(\lambda) &= [\tilde{V}^{-1}(\lambda)\tilde{H}]^{+}\tilde{V}^{-1}\tilde{z}\\ &= [F^{\mathrm{T}}F+\lambda^{-2}G^{\mathrm{T}}G]^{+}[F^{\mathrm{T}}w+\lambda^{-2}G^{\mathrm{T}}u].\end{aligned}$$

By (4.9),

$$[F^{\mathrm{T}}F+\lambda^{-2}G^{\mathrm{T}}G]^{+} = (\bar{F}^{\mathrm{T}}\bar{F})^{+} + \lambda^2(I-\bar{F}^{+}F)[(G^{+}G)^{+} + J(\lambda)](I-\bar{F}^{+}F)^{\mathrm{T}},$$

where

$$\bar{F} = F(I-G^{+}G) \qquad \text{and} \qquad J(\lambda) = O(\lambda^2) \qquad \text{as} \quad \lambda\to 0.$$

Thus

$$\begin{aligned}\tilde{x}(\lambda) &= (\bar{F}^{\mathrm{T}}\bar{F})^{+}F^{\mathrm{T}}w + \lambda^{-2}(\bar{F}^{\mathrm{T}}\bar{F})^{+}G^{\mathrm{T}}u\\ &\quad + (I-\bar{F}^{+}F)[(G^{\mathrm{T}}G)^{+} + O(\lambda^2)](I-\bar{F}^{+}F)^{\mathrm{T}}G^{\mathrm{T}}u \quad \text{as} \quad \lambda\to 0.\end{aligned} \tag{7.1.10.3}$$

But

$$\bar{F}G^{\mathrm{T}} = F(I-G^{+}G)G^{\mathrm{T}} = F[G(I-G^{+}G)]^{\mathrm{T}} = 0 \tag{7.1.10.4}$$

so

$$\mathscr{R}[G^{\mathrm{T}}] \subseteq \mathscr{N}(\bar{F}) = \mathscr{N}(\bar{F}^{+\mathrm{T}}) = \mathscr{N}(\bar{F}^{+}\bar{F}) = \mathscr{N}[(\bar{F}^{\mathrm{T}}\bar{F})^{+}] \qquad (3.11.5)$$

and so

$$(\bar{F}^{\mathrm{T}}\bar{F})^{+}G^{\mathrm{T}} = 0 \qquad \text{and} \qquad \bar{F}^{+\mathrm{T}}G^{\mathrm{T}} = 0. \tag{7.1.10.5}$$

Therefore

$$\begin{aligned}(I-\bar{F}^{+}F)(G^{\mathrm{T}}G)^{+}(I-\bar{F}^{+}F)^{\mathrm{T}}G^{\mathrm{T}} &= (I-\bar{F}^{+}F)(G^{\mathrm{T}}G)^{+}G^{\mathrm{T}}\\ &= (I-\bar{F}^{+}F)G^{+} \qquad (3.8.1)\end{aligned} \tag{7.1.10.6}$$

and by (3.13.10)

(7.1.10.7) $$(\bar{F}^{\mathrm{T}}\bar{F})^{+}F^{\mathrm{T}} = F^{+}.$$

Combining (7.1.10.3)–(7.1.10.6),

(7.1.10.8) $$\tilde{x}(\lambda) = \bar{F}^{+}w + (I-\bar{F}^{+}F)[G^{+}u+O(\lambda^2)] \qquad \text{as} \quad \lambda \to 0$$

so

(7.1.10.9) $$\tilde{x}(\lambda) = x^0 + O(\lambda^2) \qquad \text{as} \quad \lambda \to 0$$

where

(7.1.10.10) $$\begin{aligned} x^0 &= \bar{F}^{+}w + (I-\bar{F}^{+}F)\,G^{+}u \\ &= \bar{F}^{+}(w-FG^{+}u) + G^{+}u. \end{aligned}$$

By (3.12.7), the latter is exactly the vector of minimum norm which minimizes $\|w-Fx\|^2$ among those vectors which also minimize $\|u-Gx\|^2$. This proves (b) and (c), since

$$\|w-Fx\|^2 = (z-Hx)^{\mathrm{T}}V^{-2}(z-Hx).$$

(d) $$\begin{aligned} &\lim_{\lambda\to 0}[\tilde{z}-\tilde{H}\tilde{x}(\lambda)]^{\mathrm{T}}\tilde{V}^{-2}(\lambda)[\tilde{z}-\tilde{H}\tilde{x}(\lambda)] \\ &\quad = \lim_{\lambda\to 0}\{\|w-F\tilde{x}(\lambda)\|^2 + \lambda^{-2}\|u-G\tilde{x}(\lambda)\|^2\} \\ &\quad = \lim_{\lambda\to 0}\{\|w-Fx^0+O(\lambda^2)\|^2 + \lambda^{-2}\|u-Gx^0+O(\lambda^2)\|^2\}. \end{aligned}$$

If $u \in \mathscr{R}(G)$, then x^0 must satisfy the equation $Gx = u$ if it minimizes $\|Gx-u\|^2$, so that the last term tends to zero as $\lambda \to 0$ while the first tends to $\|w-Fx^0\|^2$. ■

(7.2) Constrained Least Squares Estimators as Limiting Cases of BLUE's

By (6.1.12c), $\tilde{x}(\lambda)$, defined in (7.1.10), coincides with the BLUE for x when observations of the form

(7.2.1) $$\tilde{\mathbf{z}} = \tilde{H}x + \tilde{\mathbf{v}}$$

are used to estimate x, where $\tilde{\mathbf{v}}$ is a vector random variable with mean zero and covariance $\tilde{V}^2(\lambda)$, where $\tilde{V}(\lambda)$ is defined in (7.1.10). Thus, (7.1.10) shows that any constrained weighted least squares estimator can be viewed as the limiting case of a BLUE, some of whose observations are extremely reliable (i.e., have extremely low residual variances). To put it another way, constrained, weighted least squares estimators can be approximated arbitrarily well, by treating the constraints as fictitious "observations" which are extremely accurate, and computing the BLUE for x using both the "real" observations (pretending they have covariance V) and the "fictitious" ones (pretending they have covariance $\lambda^2 I$, with λ^2 small).

(7.2.2) **Exercise:** If $\tilde{\mathbf{v}}$ has covariance

$$\tilde{V}(0) = \left(\begin{array}{c|c} V & 0 \\ \hline 0 & 0 \end{array}\right)$$

the BLUE for x is given by

(7.2.2.1) $$\tilde{\mathbf{x}}(0) = \tilde{H}^{+}[I - \tilde{V}(0)(\tilde{Q}\tilde{V}(0))^{+}]\tilde{\mathbf{z}}$$

where $\tilde{Q} = (I - \tilde{H}\tilde{H}^{+})$, (6.1.12). Is it true that $\tilde{\mathbf{x}}(0)$ coincides with $x^0 = \lim_{\lambda \to 0} \tilde{\mathbf{x}}(\lambda)$? (This would mean that constrained least squares estimates are obtainable as BLUE'S by treating the constraints as "perfectly noiseless" observations. See Zyskind and Martin [1]; Goldman and Zelen [1].)

Chapter VIII

RECURSIVE COMPUTATION OF LEAST SQUARES ESTIMATORS

(8.1) Unconstrained Least Squares

In many contemporary applications of the least squares technique, the data arrives in a stream (i.e., in temporal succession) and in such cases it is desirable to carry out the computation of the least squares estimate so that at each instant of time it fully reflects all the data that have been collected so far (cf. American Statistical Association [1] for notable applications of this philosophy).

In Chapter V, several algorithms were described for computing H^+z, but none of these were ideally suited for a situation where the least squares estimate is to be updated every time a new datum is obtained.

For the sake of concreteness, we will initially view this question in the framework of linear regression analysis, but this interpretation is not essential. We are really just relating the least squares estimate for x, based on n data points, to the least squares estimate for x based on $n+1$ data points.

Imagine a stream of scalar observations, $\zeta_1, \zeta_2, \ldots$ arriving in temporal succession, and assume that the jth observation is of the form

(8.1.1) $$\zeta_j = h_j^{\mathrm{T}} x + \mathbf{v}_j$$

where each h_j is a known p-dimensional vector, the $\mathbf{v}_j$'s are uncorrelated zero mean residuals with common variance and x is an unknown p-dimensional vector.

In (6.1.12), we showed that the BLUE for x based on the data

$$\tilde{\mathbf{z}}_n = \begin{pmatrix} \zeta_1 \\ \vdots \\ \zeta_n \end{pmatrix}$$

is

$$\hat{\mathbf{x}}_n = H_n^{+} \mathbf{z}_n \tag{8.1.2}$$

where

$$H_n = \begin{bmatrix} h_1^{\mathrm{T}} \\ \hdashline h_2^{\mathrm{T}} \\ \hdashline \vdots \\ \hdashline h_n^{\mathrm{T}} \end{bmatrix}.$$

If a new observation of the form

$$\zeta_{n+1} = h_{n+1}^{\mathrm{T}} x + \mathbf{v}_{n+1}$$

is taken, then the BLUE for x is

$$\hat{\mathbf{x}}_{n+1} = H_{n+1}^{+} \mathbf{z}_{n+1} \tag{8.1.3}$$

where

$$H_{n+1} = \begin{bmatrix} H_n \\ \hdashline h_{n+1}^{\mathrm{T}} \end{bmatrix}$$

and

$$\mathbf{z}_{n+1} = \begin{bmatrix} \mathbf{z}_n \\ \hdashline \zeta_{n+1} \end{bmatrix}.$$

It turns out that $\hat{\mathbf{x}}_{n+1}$ is nicely related to $\hat{\mathbf{x}}_n$: By (4.3.2)

$$H_{n+1}^{+} = [(I - K_{n+1} h_{n+1}^{\mathrm{T}}) H_n^{+} \mid K_{n+1}] \tag{8.1.4}$$

where

$$K_{n+1} = \begin{cases} \dfrac{(I - H_n^{+} H_n) h_{n+1}}{h_{n+1}^{\mathrm{T}} (I - H_n^{+} H_n) h_{n+1}} & \text{if } (I - H_n^{+} H_n) h_{n+1} \neq 0 \\[2ex] \dfrac{H_n^{+} H_n^{\mathrm{T}+} h_{n+1}}{1 + h_{n+1}^{\mathrm{T}} H_n^{+} H_n^{\mathrm{T}+} h_{n+1}} & \text{otherwise.} \end{cases} \tag{8.1.5}$$

Thus,

(8.1.6)
$$\begin{aligned}\hat{\mathbf{x}}_{n+1} &\equiv H_{n+1}^{+}\mathbf{z}_{n+1}\\ &= (I-K_{n+1}h_{n+1}^{\mathrm{T}})H_n{}^{+}\mathbf{z}_n + K_{n+1}\zeta_{n+1}\\ &= \hat{\mathbf{x}}_n + K_{n+1}[\zeta_{n+1}-h_{n+1}^{\mathrm{T}}\hat{\mathbf{x}}_n]\end{aligned}$$

where

$$\hat{\mathbf{x}}_0 = 0.$$

Notice that the recursion (8.1.6) takes the so-called "differential-correction" form: If $\hat{\mathbf{x}}_n$ is used as an estimate for x, a "predictor" for ζ_{n+1} (the next datum) is $h_{n+1}^{\mathrm{T}}\hat{\mathbf{x}}_n$. The prediction error is $\zeta_{n+1}-h_{n+1}^{\mathrm{T}}\hat{\mathbf{x}}_n$. From (8.1.6) the new estimate $\hat{\mathbf{x}}_{n+1}$ is obtained from $\hat{\mathbf{x}}_n$ by adding on a term proportional to the prediction error. The "vector of proportionality" K_{n+1} is sometimes referred to as the "smoothing vector." It does not depend on the data, (8.1.5).

The recursion need not be restricted in its interpretation to the present statistical domain of discourse. It is a fully general result which relates $H^{+}z$ to

$$\begin{pmatrix} H \\ \hline h^{\mathrm{T}} \end{pmatrix}^{+}\begin{pmatrix} z \\ \hline \zeta \end{pmatrix},$$

where h^{T} is an arbitrary row vector (of the correct size) and ζ is an arbitrary scalar.

(8.1.7) *Special case:* Suppose H_n is of full rank (= the number of columns of H_n). This means that H_n's columns are linearly independent. So then are H_{n+1}'s. Consequently, $H_m{}^{\mathrm{T}}H_m$ has an inverse for all $m \geqslant n$.

This allows an alternative derivation: Let

$$B_m = (H_m{}^{\mathrm{T}}H_m)^{-1} \qquad \text{for} \quad m \geqslant n.$$

We have already shown, (4.6.1), that

(8.1.7.1)
$$B_{m+1} = B_m - \frac{(B_m h_{m+1})(B_m h_{m+1})^{\mathrm{T}}}{1+h_{m+1}^{\mathrm{T}}B_m h_{m+1}}$$

where h_{m+1}^{T} is the row vector that is adjoined to H_m to produce H_{m+1}.

Since

$$\begin{aligned}\hat{\mathbf{x}}_{m+1} &= H_{m+1}^{+}\mathbf{z}_{m+1} = B_{m+1}H_{m+1}^{\mathrm{T}}\mathbf{z}_{m+1}\\ &= B_{m+1}(H_m{}^{\mathrm{T}} \vdots h_{m+1})\begin{pmatrix}\mathbf{z}_m \\ \hline \zeta_{m+1}\end{pmatrix}\end{aligned}$$

we can apply (8.1.7.1) and obtain (using the symmetry of B_m)

$$\hat{\mathbf{x}}_{m+1} = J_{m+1} B_m [H_m{}^{\mathrm{T}} \mathbf{z}_m + h_{m+1} \zeta_{m+1}]$$

where

$$J_{m+1} = I - \frac{(B_m h_{m+1}) h_{m+1}^{\mathrm{T}}}{1 + h_{m+1}^{\mathrm{T}} B_m h_{m+1}}.$$

Since $B_m H_m{}^{\mathrm{T}} \mathbf{z}_m = H_m{}^{+} \mathbf{z}_m = \hat{\mathbf{x}}_m$ and

$$J_{m+1} B_m h_{m+1} = \frac{B_m h_{m+1}}{1 + h_{m+1}^{\mathrm{T}} B_m h_{m+1}}$$

we find that

(8.1.7.2) $$\hat{\mathbf{x}}_{m+1} = \hat{\mathbf{x}}_m + \frac{B_m h_{m+1}}{1 + h_{m+1}^{\mathrm{T}} B_m h_{m+1}} [\zeta_{m+1} - h_{m+1}^{\mathrm{T}} \hat{\mathbf{x}}_m].$$

Since the columns of H_m are linearly independent by assumption, $\mathcal{N}(H_m) = \{0\}$, hence the projection on $\mathcal{N}(H_m)$ is zero: $I - H_m{}^{+} H_m = 0$. It follows that $(I - H_m{}^{+} H_m) h_{m+1} = 0$, so that K_{m+1}, defined in (8.1.5), is given by the second half of the right side which coincides with $B_m h_{m+1}/1 + h_{m+1}^{\mathrm{T}} B_m h_{m+1}$ as it should.

(8.1.8) **Exercise**

(a) h_{n+1} is a linear combination (l.c.) of $h_1, \ldots, h_n$ if and only if $(I - H_n{}^{+} H_n) h_{n+1} = 0$.

(b) Let $A_n = I - H_n{}^{+} H_n$ and $B_n = (H_n{}^{\mathrm{T}} H_n)^{+}$. Then

(8.1.8.1) $$K_{n+1} = \begin{cases} \dfrac{A_n h_{n+1}}{h_{n+1}^{\mathrm{T}} A_n h_{n+1}} & \text{if } h_{n+1} \text{ is not a l.c. of } h_1, \ldots, h_n \\[2ex] \dfrac{B_n h_{n+1}}{1 + h_{n+1}^{\mathrm{T}} B_n h_{n+1}} & \text{otherwise} \end{cases}$$

where

(8.1.8.2) $$A_{n+1} = \begin{cases} A_n - \dfrac{(A_n h_{n+1})(A_n h_{n+1})^{\mathrm{T}}}{h_{n+1}^{\mathrm{T}} A_n h_{n+1}} & \text{if } h_{n+1} \text{ is not a l.c. of } h_1, \ldots, h_n \\[2ex] A_n & \text{otherwise} \end{cases}$$

$$A_0 = I \qquad B_0 = 0$$

and

$$(8.1.8.3)\quad B_{n+1} = \begin{cases} B_n - \dfrac{(B_n h_{n+1})(A_n h_{n+1})^{\mathrm{T}} + (A_n h_{n+1})(B_n h_{n+1})^{\mathrm{T}}}{h_{n+1}^{\mathrm{T}} A_n h_{n+1}} \\ \quad + \dfrac{1+h_{n+1}^{\mathrm{T}} B_n h_{n+1}}{(h_{n+1}^{\mathrm{T}} A_n h_{n+1})^2}(A_n h_{n+1})(A_n h_{n+1})^{\mathrm{T}} & \text{if } h_{n+1} \text{ is not a l.c. of } h_1, \ldots, h_n \\ B_n - \dfrac{(B_n h_{n+1})(B_n h_{n+1})^{\mathrm{T}}}{1+h_{n+1}^{\mathrm{T}} B_n h_{n+1}} & \text{otherwise.} \end{cases}$$

[*Hint:* Use (4.6) and (3.14.1) and/or (4.6.4d).]

(8.2) Recursive Computation of Residual Errors

Denote the residual error associated with $x_n = H_n^{+} z_n$, by ε_n:

$$(8.2.1)\qquad \varepsilon_n = \|z_n - H_n \hat{x}_n\|^2.$$

The results of (8.1) can be used to relate ε_{n+1} and ε_n:

(8.2.2) **Theorem:** $\varepsilon_0 = 0$

$$\varepsilon_{n+1} = \begin{cases} \varepsilon_n & \text{if } h_{n+1} \text{ is not a l.c. of } h_1, \ldots, h_n \\ \varepsilon_n + \dfrac{(\zeta_{n+1} - h_{n+1}^{\mathrm{T}} \hat{x}_n)^2}{1+h_{n+1}^{\mathrm{T}} B_n h_{n+1}} & \text{otherwise} \end{cases}$$

where $B_n = (H_n^{\mathrm{T}} H_n)^+$ satisfies (8.1.8.3).

Comment: (8.2.1) is the same as $\sum_{j=1}^{n} (\zeta_j - h_j^{\mathrm{T}} \hat{x}_n)^2$, so (8.2.2) asserts the truth of the following identity:

$$\sum_{j=1}^{n} (\zeta_j - h_j^{\mathrm{T}} \hat{x}_n)^2 = \sum_{j \in J_n} (\zeta_j - h_j^{\mathrm{T}} \hat{x}_{j-1})^2/(1+h_j^{\mathrm{T}} B_{j-1} h_j)$$

where

$$J_n = \{j : j \leqslant n \quad \text{and} \quad h_j \text{ is a l.c. of } h_1, \ldots, h_{j-1}\}.$$

The identity holds for all scalars $\zeta_1, \ldots, \zeta_n$, all vectors, $h_1, h_2, \ldots, h_n$, and all n.

Proof: Since

$$H_{n+1} = \begin{bmatrix} H_n \\ \hline h_{n+1}^{\mathrm{T}} \end{bmatrix} \quad \text{and} \quad z_{n+1} = \begin{bmatrix} z_n \\ \hline \zeta_{n+1} \end{bmatrix}$$

we see from (8.1.6) that

(8.2.2.1)

$$z_{n+1} - H_{n+1} \hat{x}_{n+1} = \left(\frac{z_n - H_n \hat{x}_n}{\zeta_{n+1} - h_{n+1}^{\mathrm{T}} \hat{x}_n}\right) - \left(\frac{H_n K_{n+1}}{h_{n+1}^{\mathrm{T}} K_{n+1}}\right)(\xi_{n+1} - h_{n+1}^{\mathrm{T}} \hat{x}_n).$$

Case 1: h_{n+1} is not a l.c. of $h_1, \ldots, h_n$:
Then $(I-H_n{}^+H_n)h_{n+1} \neq 0$, and K_{n+1} is defined by the first part of (8.1.8.1). Therefore

$$H_n K_{n+1} = H_n A_n h_{n+1}/(h_{n+1}^{\mathrm{T}} A_n h_{n+1}).$$

Since

$$A_n = I - H_n{}^+ H_n$$

it follows that

$$H_n A_n = 0$$

so

$$H_n K_{n+1} = 0 \qquad \text{and} \qquad h_{n+1}^{\mathrm{T}} K_{n+1} = 1.$$

Hence by (8.2.2.1),

$$\begin{aligned} \varepsilon_{n+1} &= \|z_{n+1} - H_{n+1}\hat{x}_{n+1}\|^2 \\ &= \|z_n - H_n \hat{x}_n\|^2 = \varepsilon_n. \end{aligned}$$

Case 2: h_{n+1} is a l.c. of $h_1, \ldots, h_n$: In this case, K_n is defined by the second half of (8.1.8.1). Then by (8.2.2.1),

(8.2.2.2) $$\begin{aligned} &\|z_{n+1} - H_{n+1}\hat{x}_{n+1}\|^2 \\ &= \|z_n - H_n\hat{x}_n\|^2 + (\zeta_{n+1} - h_{n+1}^{\mathrm{T}}\hat{x}_n)^2 \\ &\quad - 2\{(z_n - H_n\hat{x}_n)^{\mathrm{T}} H_n K_{n+1} + (\zeta_{n+1} - h_{n+1}^{\mathrm{T}}\hat{x}_n) h_{n+1}^{\mathrm{T}} K_{n+1}\} \\ &\quad \times (\zeta_{n+1} - h_{n+1}^{\mathrm{T}}\hat{x}_n) \\ &\quad + \{K_{n+1}^{\mathrm{T}} H_n{}^{\mathrm{T}} H_n K_{n+1} + (h_{n+1}^{\mathrm{T}} K_{n+1})^2\}(\zeta_{n+1} - h_{n+1}^{\mathrm{T}}\hat{x}_n)^2 \\ &= \|z_n - H_n\hat{x}_n\|^2 - 2z_n{}^{\mathrm{T}}(I - H_n H_n{}^+) H_n K_{n+1}(\zeta_{n+1} - h_{n+1}^{\mathrm{T}}\hat{x}_n) \\ &\quad + [(K_{n+1}^{\mathrm{T}} H_n{}^{\mathrm{T}} H_n K_{n+1}) + (1 - h_{n+1}^{\mathrm{T}} K_{n+1})^2](\zeta_{n+1} - h_{n+1}^{\mathrm{T}}\hat{x}_n)^2. \end{aligned}$$

[We have used the fact that $(z_n - H_n\hat{x}_n)^{\mathrm{T}} = z^{\mathrm{T}}(I - H_n H_n{}^+)$ to generate the second term in (8.2.2.2.).]

Since $(I - H_n H_n{}^+) H_n = 0$, the second term vanishes. Since K_{n+1} is defined by the second half of (8.1.8.1) and since

$$B_n{}^{\mathrm{T}}(H_n{}^{\mathrm{T}} H_n) B_n = B_n$$

[because $B_n = (H_n{}^{\mathrm{T}} H_n)^+$, which is symmetric]
it follows that

$$K_{n+1}^{\mathrm{T}}(H_n{}^{\mathrm{T}} H_n) K_{n+1} = \frac{h_{n+1}^{\mathrm{T}} B_n h_{n+1}}{(1 + h_{n+1}^{\mathrm{T}} B_n h_{n+1})^2}$$

and

$$(1-h_{n+1}^{\mathrm{T}}K_{n+1})^2 = \frac{1}{(1+h_{n+1}^{\mathrm{T}}B_n h_{n+1})^2}.$$

The third term in (8.2.2.2), reduces to $(\zeta_{n+1}-h_{n+1}^{\mathrm{T}}\hat{x}_n)^2/(1+h_{n+1}^{\mathrm{T}}B_n h_{n+1})$ and this proves the theorem. ■

(8.3) Weighted Least Squares

The recursion (8.1.6) and (8.1.8) can be viewed as a relationship between the value of x which minimizes

$$\sum_{j=1}^{n}(\zeta_j-h_j^{\mathrm{T}}x)^2$$

and the value of x which minimizes

$$\sum_{j=1}^{n+1}(\zeta_j-h_j^{\mathrm{T}}x)^2.$$

Not surprisingly, a similar relationship holds between the values of x which respectively minimize

$$\sum_{j=1}^{n}(\zeta_j-h_j^{\mathrm{T}}x)^2/\sigma_j^2 \qquad \text{and} \qquad \sum_{j=1}^{n+1}(\zeta_j-h_j^{\mathrm{T}}x)/\sigma_j^2.$$

(8.3.1) **Exercise:** Let $\sigma_1^2, \sigma_2^2, \ldots$ be a sequence of positive scalars and let $\tilde{x}_n$ be the vector of minimum norm among those which minimize $\sum_{j=1}^{n}(\zeta_j-h_j^{\mathrm{T}}x)/\sigma_j^2$. Then $\tilde{x}_0=0$,

(8.3.1.1) $$\tilde{x}_{n+1} = \tilde{x}_n + \tilde{K}_{n+1}(\zeta_{n+1}-h_{n+1}^{\mathrm{T}}\tilde{x}_n)$$

where

(8.3.1.2) $$\tilde{K}_{n+1} = \begin{cases} A_n h_{n+1}/h_{n+1}^{\mathrm{T}}A_n h_{n+1} & \text{if } h_{n+1} \text{ is not a l.c. of } h_1,\ldots,h_n \\ \tilde{B}_n h_{n+1}/(\sigma_{n+1}^2+h_{n+1}^{\mathrm{T}}\tilde{B}_n h_{n+1}) & \text{otherwise,} \end{cases}$$

$$A_0 = I,$$

(8.3.1.3) $$A_{n+1} = \begin{cases} A_n - \dfrac{(A_n h_{n+1})(A_n h_{n+1})^{\mathrm{T}}}{h_{n+1}^{\mathrm{T}}A_n h_{n+1}} & \text{if } h_{n+1} \text{ is not a l.c. of } h_1,\ldots,h_n \\ A_n & \text{otherwise,} \end{cases}$$

$$\tilde{B}_0 = 0,$$

$$(8.3.1.4)\quad \tilde{B}_{n+1} = \begin{cases} \tilde{B}_n - \dfrac{(\tilde{B}_n h_{n+1})(A_n h_{n+1})^{\mathrm{T}} + (A_n h_{n+1})(\tilde{B}_n h_{n+1})^{\mathrm{T}}}{h_{n+1}^{\mathrm{T}} A_n h_{n+1}} \\ \quad + \dfrac{\sigma_{n+1}^2 + h_{n+1}^{\mathrm{T}} \tilde{B}_n h_{n+1}}{(h_{n+1}^{\mathrm{T}} A_n h_{n+1})^2} (A_n h_{n+1})(A_n h_{n+1})^{\mathrm{T}} & \text{if } h_{n+1} \text{ is not a l.c. of } h_1, \ldots, h_n \\ \tilde{B}_n - \dfrac{(\tilde{B}_n h_{n+1})(\tilde{B}_n h_{n+1})^{\mathrm{T}}}{\sigma_{n+1}^2 + h_{n+1}^{\mathrm{T}} \tilde{B}_n h_{n+1}} & \text{otherwise.} \end{cases}$$

Denoting the associated residual error by $\tilde{\varepsilon}_n (= \sum_{j=1}^n (\zeta_j - h_j^{\mathrm{T}} \tilde{x}_n)^2/\sigma_j^2)$, we have

$$\tilde{\varepsilon}_0 = 0,$$

$$(8.3.1.5)\quad \tilde{\varepsilon}_{n+1} = \begin{cases} \tilde{\varepsilon}_n & \text{if } h_{n+1} \text{ is not a l.c. of } h_1, \ldots, h_n \\ \tilde{\varepsilon}_n + \dfrac{(\zeta_{n+1} - h_{n+1}^{\mathrm{T}} \tilde{x}_n)^2}{(\sigma_{n+1}^{\mathrm{T}} + h_{n+1}^2 \tilde{B}_n h_{n+1})} & \text{otherwise.} \end{cases}$$

For each n,

$$(8.3.1.6)\qquad A_n = I - H_n^{+} H_n$$

and

$$(8.3.1.7)\qquad \tilde{B}_n = \left(\sum_{j=1}^n h_j h_j^{\mathrm{T}}/\sigma_j^2 \right)^{+} = (H_n^{\mathrm{T}} V_n^{-2} H_n)^{+}$$

where

$$(8.3.1.8)\qquad H_n = \begin{pmatrix} h_1^{\mathrm{T}} \\ \vdots \\ h_n^{\mathrm{T}} \end{pmatrix} \qquad \text{and} \qquad V_n^2 = \operatorname{diag}(\sigma_1^2, \ldots, \sigma_n^2).$$

[*Hint:* Let $\tilde{\zeta}_j = \zeta_j/\sigma_j$, $\tilde{h}_j = h_j/\sigma_j$ and notice that

$$\mathscr{L}(h_1, h_2, \ldots, h_n) = \mathscr{L}(\tilde{h}_1, \tilde{h}_2, \ldots, \tilde{h}_n)$$

and

$$\sum_{j=1}^n (\zeta_j - h_j^{\mathrm{T}} x)^2/\sigma_j^2 = \sum_{j=1}^n (\tilde{\zeta}_j - \tilde{h}_j^{\mathrm{T}} x)^2.$$

Apply (8.1.6)–(8.1.8) to the $\tilde{\zeta}_j$ and $\tilde{h}_j$'s and then translate back to ζ_j's and h_j's.]

(8.4) Recursive Constrained Least Squares, I

In (3.12.4b) we showed that

(8.4.1) $$x^0 = G^+u + [H(I-G^+G)]^+(z-HG^+u)$$

is the vector of minimum norm among those which minimize

(8.4.2) $$\|z-Hx\|^2$$

subject to the constraints

(8.4.3) $$Gx = u$$

provided the constraint set is nonempty.

Imagine a "stream" of data, $\zeta_1, \zeta_2, \ldots, \zeta_n, \ldots$, and denote the vector of n observations by z_n and let $x_n{}^0$ be the minimum norm vector among those which minimize $\|z_n - H_n x\|^2 = \sum_{j=1}^n (\zeta_j - h_j{}^\mathrm{T} x)^2$, subject to $Gx = 0$:

(8.4.4) $$x_n{}^0 = G^+u + \bar{H}_n{}^+ \bar{z}_n$$

where $\bar{H}_n = H_n(I-G^+G)$ is the matrix whose jth row is

$$\bar{h}_j{}^\mathrm{T} = [(I-G^+G)h_j]^\mathrm{T} \qquad (j = 1, 2, \ldots, n)$$

and $\bar{z}_n = z_n - H_n G^+ u$ is the n vector whose jth component is

$$\bar{\zeta}_j = \zeta_j - h_j{}^\mathrm{T} G^+ u \qquad (j = 1, 2, \ldots, n).$$

The recursion for $\hat{x}_n = H_n{}^+ z_n$ can be applied almost verbatim to the computation of

$$x_n{}^0 = G^+u + \bar{H}^+ \bar{z}_n.$$

(8.4.5) **Theorem:** Let $x_0{}^0 = G^+u$, $x_{n+1}^0 = x_n{}^0 + \bar{K}_{n+1}(\bar{\zeta}_{n+1} - \bar{h}_{n+1}^\mathrm{T} x_n{}^0)$ where

$$\bar{K}_{n+1} = \begin{cases} \dfrac{\bar{A}_n \bar{h}_{n+1}}{\bar{h}_{n+1}^\mathrm{T} \bar{A}_n \bar{h}_{n+1}} & \text{if } \bar{h}_{n+1} \text{ is not a l.c. of } \bar{h}_1, \bar{h}_2, \ldots, \bar{h}_n \\[2ex] \dfrac{\bar{B}_n \bar{h}_{n+1}}{(1 + \bar{h}_{n+1}^\mathrm{T} \bar{B}_n \bar{h}_{n+1})} & \text{otherwise} \end{cases}$$

where

$$\bar{A}_n = I - \bar{H}_n{}^+ \bar{H}_n \qquad \text{and} \qquad \bar{B}_n = (\bar{H}_n{}^\mathrm{T} \bar{H}_n)^+$$

can be computed recursively by (8.1.8.2) and (8.1.8.3) with h's replaced by $\bar{h}$'s throughout.

Proof: Let $y_n{}^0$ satisfy the same recursion as above except that $y_0{}^0 = 0$. Then by (8.1.6), $y_n{}^0 = \bar{H}_n{}^+ \bar{z}_n$ for every n. If we set $d_n = x_n{}^0 - y_n{}^0$, then d_n satisfies

$$d_{n+1} = (I - \bar{K}_{n+1} \bar{h}_{n+1}^{\mathrm{T}}) d_n = \left[\prod_{j=1}^{n+1} (I - \bar{K}_j \bar{h}_j{}^{\mathrm{T}}) \right] d_0.$$

But $d_0 = x_0{}^0 = G^+ u$ and $\bar{h}_j{}^{\mathrm{T}} d_0 = h_j{}^{\mathrm{T}}(I - G^+ G) G^+ u = 0$ so that

$$d_{n+1} = d_0 \qquad \text{for all} \quad n$$

which proves that $x_{n+1}^0 = y_{n+1}^0 + d_0$: Thus,

$$x_{n+1}^0 = G^+ u + \bar{H}_n{}^+ \bar{z}_n. \quad \blacksquare$$

If the previously mentioned recursion is used to process data in "real time," one precomputes $G^+ u$ and $I - G^+ G$. As each datum ζ_j is obtained, it is transformed to $\bar{\zeta}_j = \zeta_j - h_j{}^{\mathrm{T}} G^+ u$ and the corresponding regression vector is transformed to $\bar{h}_j = (I - G^+ G) h_j$. The $\bar{A}_n$'s and $\bar{B}_n$'s are computed by the ancillary recursions (8.1.8.2) and (8.1.8.3).

The vector $G^+ u$ can be computed by any of the methods of Chapter V or recursively: Let the rows of G be denoted by $g_1{}^{\mathrm{T}}, \ldots, g_k{}^{\mathrm{T}}$. Define

$$G_m = \begin{pmatrix} g_1{}^{\mathrm{T}} \\ \vdots \\ g_m{}^{\mathrm{T}} \end{pmatrix} \qquad m = 1, 2, \ldots, k.$$

Use (8.1.6), (8.1.8.2), and (8.1.8.3) to compute $G_m{}^+ u_m$ where u_m is the vector consisting of u's first m component, for $m = 1, 2, \ldots, k$. As a by-product, the A recursion will generate $I - G^+ G$.

(8.4.6) **Exercise:** Let

$$G = \begin{pmatrix} g_1{}^{\mathrm{T}} \\ \vdots \\ g_k{}^{\mathrm{T}} \end{pmatrix}$$

and let $h_1, \ldots, h_n, \ldots$ be a sequence of vectors having the same dimensionality as the g's. Let

$$\bar{h}_j = (I - G^+ G) h_j \qquad (j = 1, 2, \ldots).$$

Show that for any n, $\bar{h}_n$ is a linear combination of $\bar{h}_1, \bar{h}_2, \ldots, \bar{h}_{n-1}$ if and only if

$$h_n \in \mathscr{L}(g_1, \ldots, g_k, h_1, \ldots, h_{n-1})$$

(i.e., h_n is a linear combination of $h_1, \ldots, h_{n-1}$ and the g's).

(8.4.7) **Exercise:** Suppose $f \in \mathscr{L}(g_1, \ldots, g_k, h_1, \ldots, h_n)$: Then

$$\lim_{\lambda \to 0} \left(\sum_{j=1}^{n} h_j h_j^{\mathrm{T}} + \lambda^{-2} \sum_{j=1}^{k} g_j g_j^{\mathrm{T}} \right)^{+} f = 0$$

if and only if $f \in \mathscr{L}(g_1, \ldots g_k)$.

(8.4.8) **Exercise:** Extend the results of (8.4.5) and develop a recursion for $\tilde{x}_n^{\,0}$, the value of x which minimizes

$$\sum_{j=1}^{n} (\zeta_j - h_j^{\mathrm{T}} x)/\sigma_j^2$$

subject to $Gx = u$.

(8.5) Recursive Constrained Least Squares, II

In (8.4) we showed how constraints can be incorporated into a least squares recursion by altering the h-vectors and the observations, and by suitably initializing the x-vector.

An essential feature of the technique is the necessity to specify the constraints in *advance* of the data acquisition. Under certain circumstances, this is a drawback. Often, one wishes to collect all the data and afterward compute the least squares estimate subject to a succession of progressively more restrictive linear side conditions. By examining the residual errors [cf. (6.4)] one can judge the reasonableness of the constraints in the light of the available data. Or, one may wish to incorporate a few side conditions into the computation at the outset and then, after the data acquisition (or perhaps, even *during*) experiment with constraints.

In this section we shall show that constraints can be incorporated into the recursion (8.3.1) by treating them as fictitious observations with zero variance.

(8.5.1) **Theorem:** Let $h_1, h_2, \ldots$ be a sequence of vectors, let $\zeta_1, \zeta_2, \ldots$ be a sequence of scalars, and let $\sigma_1^2, \sigma_2^2, \ldots$ be another sequence of nonnegative (but not necessarily positive) scalars. Let $\tilde{x}_n$, $\tilde{K}_n$, $\tilde{B}_n$, A_n, and $\tilde{\varepsilon}_n$ be defined recursively as in (8.3.1).

(a) If the σ_j^2 are all positive, then for each n, $\tilde{x}_n$ minimizes the weighted sum of squares

$$\sum_{j=1}^{n} (\zeta_j - h_j^{\mathrm{T}} x)^2/\sigma_j^2$$

$$\tilde{B}_n = \left(\sum_{j=1}^{n} h_j h_j^{\mathrm{T}}/\sigma_j^2 \right)^{+}$$

and

$$\tilde{\varepsilon}_n = \sum_{j=1}^{n} (\zeta_j - h_j^{\mathrm{T}} \tilde{x}_n)^2 / \sigma_j^2.$$

(b) If some of the σ_j^2 are zero, say

$$\sigma_j^2 \begin{cases} = 0 & \text{if } j \in S \\ > 0 & \text{if } j \notin S \end{cases}$$

where S is a finite set of integers, then for each n, $\tilde{x}_n$ minimizes

$$\sum_{j \in T_n} (\zeta_j - h_j^{\mathrm{T}} x)^2 / \sigma_j^2$$

subject to the constraints

$$h_j^{\mathrm{T}} x = \zeta_j \qquad j \in S_n$$

and

$$\tilde{\varepsilon}_n = \sum_{j \in T_n} (\zeta_j - h_j^{\mathrm{T}} \tilde{x}_n)^2 / \sigma_j^2$$

provided $\{h_j; j \in S\}$ is a linearly independent set. (Here we define $S_n = \{1, 2, \ldots, n\} \cap S$ and $T_n = S - S_n$.)

Proof: (a) We have merely restated the results of (8.3.1) for easy reference.

(b) If some of the σ_j's are zero, (8.3.1) does not apply directly, since those results were derived, under the assumption of positive σ_j's. For this reason, we define

$$\tau_j^2(\lambda) = \begin{cases} \sigma_j^2 & \text{if } j \notin S \\ \lambda^2 & \text{if } j \in S \end{cases}$$

and consider the recursions of (8.3.1) with σ_j^2 replaced throughout by $\tau_j^2(\lambda)$. The A recursion does not depend on λ, but the x, B, K, and ε recursions do, so we display the relationship by denoting the outputs of these recursions by $\tilde{x}_n(\lambda)$, $\tilde{B}_n(\lambda)$, $\tilde{K}_n(\lambda)$, and $\tilde{\varepsilon}_n(\lambda)$. We will show that $\tilde{x}_n(\lambda)$ and $\tilde{\varepsilon}_n(\lambda)$ converge to the desired quantities[1] as $\lambda \to 0$ and that $\lim_{\lambda \to 0} \tilde{x}_n(\lambda)$ and $\lim_{\lambda \to 0} \tilde{\varepsilon}_n(\lambda)$ satisfy the stated recursions with $\tilde{B}_n(\lambda)$ replaced by $\tilde{B}_n = \lim_{\lambda \to 0} \tilde{B}_n(\lambda)$, $\tilde{K}_n(\lambda)$ replaced by $\lim_{\lambda \to 0} \tilde{K}_n(\lambda)$ and $\tau_j^2(\lambda)$ replaced throughout by σ_j^2 or 0 depending on whether $j \in T_j$ or $j \in S_j$.

(1) *Convergence of $\tilde{x}_n(\lambda)$ and $\tilde{\varepsilon}_n(\lambda)$*

By (8.3.1), $\tilde{x}_n(\lambda)$ is the vector of minimum norm among those which minimize

$$\sum_{j=1}^{n} (\zeta_j - h_j^{\mathrm{T}} x)^2 / \tau_j^2(\lambda).$$

[1] The constrained least squares estimate and the associated residual error.

Therefore

(8.5.1.1) $$\tilde{x}_n(\lambda) = \tilde{H}_n^{+}(\lambda)\tilde{z}_n(\lambda) = [\tilde{H}_n^{\mathrm{T}}(\lambda)\tilde{H}_n(\lambda)]^{+}\tilde{H}_n^{\mathrm{T}}(\lambda)\tilde{z}_n(\lambda)$$

where

$$\tilde{H}_n(\lambda) = \begin{bmatrix} h_1^{\mathrm{T}}/\tau_1(\lambda) \\ \vdots \\ h_n^{\mathrm{T}}/\tau_n(\lambda) \end{bmatrix} \quad \text{and} \quad \tilde{z}_n(\lambda) = \begin{bmatrix} \zeta_1/\tau_1(\lambda) \\ \vdots \\ \zeta_n/\tau_n(\lambda) \end{bmatrix}.$$

Consequently we can rewrite (8.5.1.1) as

(8.5.1.2) $$\tilde{x}_n(\lambda) = \left[\sum_{j\in T_n} h_j h_j^{\mathrm{T}}/\sigma_j^2 + \sum_{j\in S_n} h_j h_j^{\mathrm{T}}/\lambda^2\right]^{+} \times \left(\sum_{j\in T_n} h_j^{\mathrm{T}}\zeta_j/\sigma_j^2 + \lambda^{-2}\sum_{j\in S_n} h_j^{\mathrm{T}}\zeta_j\right).$$

In (7.1.10.2)–(7.1.10.10) we showed that if the equation $Gx = u$ has a solution, and

$$\tilde{x}(\lambda) = (F^{\mathrm{T}}F+\lambda^{-2}G^{\mathrm{T}}G)^{+}(F^{\mathrm{T}}w+\lambda^{-2}G^{\mathrm{T}}u)$$

then

$$\lim_{\lambda\to 0}\tilde{x}(\lambda) = x^0$$

the vector of minimum norm among those which minimize $\|w-Fx\|^2$ subject to $Gx = u$, and that

$$\lim_{\lambda\to 0}[\|w-F\tilde{x}(\lambda)\|^2 + \lambda^{-2}\|u-G\tilde{x}(\lambda)\|^2] = \|w-Fx^0\|^2.$$

In the case at hand we let

(8.5.1.3) F be the matrix whose rows are $\{h_j^{\mathrm{T}}/\sigma_j; j\in T_n\}$

and

(8.5.1.4) G be the matrix whose rows are $\{h_j^{\mathrm{T}}; j\in S_n\}$.

(8.5.1.5) w and u are the vectors whose respective components are

$$\{\zeta_j/\sigma_j; j\in T_n\} \quad \text{and} \quad \{\zeta_j; j\in S_n\}.$$

Then (8.5.1.2) takes the form

$$\tilde{x}_n(\lambda) = [F^{\mathrm{T}}F+\lambda^{-2}G^{\mathrm{T}}G]^{+}[F^{\mathrm{T}}w+\lambda^{-2}G^{\mathrm{T}}u].$$

The rows of G are linearly independent by assumption, hence the equation $Gx = u$ has a solution, (3.12.5), hence

$$\lim_{\lambda\to 0}\tilde{x}_n(\lambda) = \tilde{x}_n$$

which minimizes $\sum_{j\in T_n}(\zeta_j-h_j^{\mathrm{T}}x)^2/\sigma_j^2$ subject to the constraints $h_j^{\mathrm{T}}x=\zeta_j$; $j\in S_n$. Furthermore

$$\lim_{\lambda\to 0}\tilde{\varepsilon}_n(\lambda)=\tilde{\varepsilon}_n=\sum_{j\in T_n}(\zeta_j-h_j^{\mathrm{T}}\tilde{x}_n)^2/\sigma_j^2.$$

(2) *Convergence of* $\tilde{B}_n(\lambda)$

$\tilde{B}_n(\lambda)=[\tilde{H}_n^{\mathrm{T}}(\lambda)\tilde{H}_n(\lambda)]^+$, (8.3.1.7), which is of the form $[F^{\mathrm{T}}F+\lambda^{-2}G^{\mathrm{T}}G]^+$ where F and G were defined earlier. By (4.9), $\lim_{\lambda\to 0}(F^{\mathrm{T}}F+\lambda^{-2}G^{\mathrm{T}}G)^+$ always exists. So then does

(8.5.1.6)
$$\tilde{B}_n=\lim_{\lambda\to 0}\tilde{B}_n(\lambda).$$

$\tilde{B}_{n+1}(\lambda)$ is related to $\tilde{B}_n(\lambda)$ through the recursion (8.3.1.4) with σ_{n+1}^2 replaced by $\tau_{n+1}^2(\lambda)$. If $n+1\in T_{n+1}$, then $\lim_{\lambda\to 0}\tau_{n+1}^2(\lambda)=\sigma_{n+1}^2$ and so $\tilde{B}_{n+1}$ is related to $\tilde{B}_n$ through (8.3.1.4). If $(n+1)\in S_{n+1}$, then $\lim_{\lambda\to 0}\tau_{n+1}^2(\lambda)=0$. $\tilde{B}_{n+1}$ will be related to $\tilde{B}_n$ through (8.3.1.4) with σ_{n+1}^2 set equal to zero if we can rule out the possibility that $h_{n+1}^{\mathrm{T}}\tilde{B}_{n+1}h_{n+1}=0$ when h_{n+1} is a l.c. of $h_1,\ldots,h_n$. But if $n+1\in S_{n+1}$ and h_{n+1} is a l.c. of $h_1,\ldots,h_n$, then $h_{n+1}^{\mathrm{T}}\tilde{B}_{n+1}h_{n+1}=0$ if and only if

$$\tilde{B}_{n+1}h_{n+1}=\lim_{\lambda\to 0}\left[\sum_{j\in T_n}h_jh_j^{\mathrm{T}}/\sigma_j^2+\lambda^{-2}\sum_{j\in S_n}h_jh_j^{\mathrm{T}}\right]^+h_{n+1}=0$$

and by (8.4.7) this can only happen if h_{n+1} is a l.c. of $\{h_j;j\in S_n\}$. The last possibility is excluded because $\{h_j;j\in S\}$ is required to be a linearly independent set by assumption. Thus

$$\tilde{B}_n=\lim_{\lambda\to 0}\tilde{B}_n(\lambda)$$

exists for each n and satisfies (8.3.1.4).

(3) *Recursions for* $\tilde{x}_n$ *and* $\tilde{\varepsilon}_n$

Since

(8.5.1.7)
$$\tilde{x}_{n+1}(\lambda)=\tilde{x}_n(\lambda)+\tilde{K}_{n+1}(\lambda)[\zeta_{n+1}-h_{n+1}^{\mathrm{T}}\tilde{x}_n(\lambda)]$$

and since

$$\tilde{K}_{n+1}=\lim_{\lambda\to 0}\tilde{K}_{n+1}(\lambda)=\begin{cases}\dfrac{A_nh_{n+1}}{h_{n+1}^{\mathrm{T}}A_nh_{n+1}} & \text{if } h_{n+1} \text{ is not a l.c. of } h_1,\ldots,h_n\\[2ex] \dfrac{\tilde{B}_nh_{n+1}}{(\sigma_{n+1}^2+h_{n+1}^{\mathrm{T}}\tilde{B}_nh_{n+1})} & \text{otherwise}\end{cases}$$

(where $\sigma_{n+1}^2=0$ if $n+1\in S$), we see that

$$\tilde{x}_{n+1}=\tilde{x}_n+\tilde{K}_{n+1}[\zeta_{n+1}-h_{n+1}^{\mathrm{T}}\tilde{x}_n]$$

since all terms on the right side of (8.5.1.7) have limits.

Similarly

$$\tilde{\varepsilon}_{n+1}(\lambda) = \tilde{\varepsilon}_n(\lambda) + \begin{cases} 0 & \text{if } h_{n+1} \text{ is not a l.c. of } h_1, \dots, h_n \\ \dfrac{[\zeta_{n+1} - h_{n+1}^{\mathrm{T}} \tilde{x}_n(\lambda)]^2}{[\tau_{n+1}^2(\lambda) + h_{n+1}^{\mathrm{T}} \tilde{B}_n(\lambda) h_{n+1}]} & \text{otherwise} \end{cases}$$

and since both sides tend to limits,

$$\tilde{\varepsilon}_{n+1} = \tilde{\varepsilon}_n + \begin{cases} 0 & \text{if } h_{n+1} \text{ is not a l.c. of } h_1, \dots, h_n \\ \dfrac{(\zeta_{n+1} - h_{n+1}^{\mathrm{T}} \tilde{x}_n)^2}{(\sigma_{n+1}^2 + h_{n+1}^{\mathrm{T}} \tilde{B}_n h_{n+1})} & \text{otherwise.} \end{cases}$$

Notice that $h_{n+1}^{\mathrm{T}} \tilde{B}_n h_{n+1} > 0$ if h_{n+1} is a l.c. of $h_1, \dots, h_n$ and $n+1 \in S_{n+1}$ (by previous argument).

(8.5.2) **Exercise:** Let $\tilde{B}_n$ be as defined earlier, and let P_n be the projection on the linear manifold spanned by $\{h_j; j \in S_n\}$. Assuming the last to be a linearly independent set, show that

$$\tilde{B}_n = \left[(I - P_n)\left(\sum_{j \in T_n} h_j h_j^{\mathrm{T}}/\sigma_j^2\right)(I - P_n)\right]^+.$$

(8.5.3) **Exercise:** If the rows of H are linearly independent of the rows of G, then

$$\min_x \|z - Hx\|^2 = \min_{Gx = u} \|z - Hx\|^2$$

(i.e., the minimal residual error for the unconstrained l.s.e. is the same as the residual error for the constrained l.s.e. in this case.)

(8.5.4) **Exercise:** Suppose the set $\{h_j; j \in S\}$ is not linearly independent. Show that

$$\tilde{B}_{n+1} = \lim_{\lambda \to 0} \tilde{B}_{n+1}(\lambda) = \tilde{B}_n$$

if h_{n+1} is a l.c. of $\{h_j; j \in S_n\}$ and $\sigma_{n+1}^2 = 0$.

(8.5.5) (*Continued*): Let $\mathscr{C}_n = \{x: h_j^{\mathrm{T}} x = \zeta_j; j \in S_n\}$.

(a) Show that

$$\lim_{\lambda \to 0} [\zeta_{n+1} - h_{n+1}^{\mathrm{T}} \tilde{x}_n(\lambda)]^2 / [\tau_{n+1}^2(\lambda) + h_{n+1}^{\mathrm{T}} \tilde{B}_n(\lambda) h_{n+1}] = 0$$

if $n+1 \in S_{n+1}$, h_{n+1} is a l.c. of $\{h_j; j \in S_n\}$, and $\mathscr{C}_{n+1}$ is not empty.

Hence, if the restriction in (8.5.1) that $\{h_j; j \in S\}$ be a linearly independent set is dropped, and if we define

$$\tilde{x}_{n+1} = \tilde{x}_n + \tilde{K}_{n+1}(\zeta_{n+1} - h_{n+1}^{\mathrm{T}}\tilde{x}_n)$$

where

$$\tilde{K}_{n+1} = \begin{cases} \dfrac{A_n h_{n+1}}{h_{n+1}^{\mathrm{T}} A_n h_{n+1}} & \text{if } h_{n+1} \text{ is not a l.c. of } h_1, \ldots, h_n \\[2ex] \dfrac{\tilde{B}_n h_{n+1}}{\sigma_{n+1}^2 + h_{n+1}^{\mathrm{T}} \tilde{B}_n h_{n+1}} & \text{otherwise} \end{cases}$$

$$\tilde{\varepsilon}_{n+1} = \tilde{\varepsilon}_n + \begin{cases} 0 & \text{if } h_{n+1} \text{ is not a l.c. of } h_1, \ldots, h_n \text{ or if } n+1 \in S_{n+1} \text{ and } h_{n+1} \text{ is a l.c. of } \{h_j; j \in S_n\} \\[2ex] \dfrac{[\zeta_{n+1} - h_{n+1}^{\mathrm{T}}\tilde{x}_n]^2}{(\sigma_{n+1}^2 + h_{n+1}^{\mathrm{T}} \tilde{B}_n h_{n+1})} & \text{otherwise,} \end{cases}$$

$$A_{n+1} = A_n - \begin{cases} \dfrac{(A_n h_{n+1})(A_n h_{n+1})^{\mathrm{T}}}{h_{n+1}^{\mathrm{T}} A_n h_{n+1}} & \text{if } h_{n+1} \text{ is a l.c. of } h_1, \ldots, h_n \\[2ex] 0 & \text{otherwise} \end{cases}$$

$$\tilde{B}_{n+1} = \begin{cases} \tilde{B}_n - \dfrac{(\tilde{B}_n h_{n+1})(A_n h_{n+1})^{\mathrm{T}} + (A_n h_{n+1})(\tilde{B}_n h_{n+1})^{\mathrm{T}}}{h_{n+1}^{\mathrm{T}} A_n h_{n+1}} \\[2ex] \quad + \dfrac{\sigma_{n+1}^2 + h_{n+1}^{\mathrm{T}} \tilde{B}_n h_{n+1}}{(h_{n+1}^{\mathrm{T}} A_n h_{n+1})^2} (A_n h_{n+1})(A_n h_{n+1})^{\mathrm{T}} \\ \qquad \text{if } h_{n+1} \text{ is not a l.c. of } h_1, \ldots, h_n \\[2ex] \tilde{B}_n \qquad \text{if } n+1 \in S_{n+1} \text{ and } h_{n+1} \text{ is a l.c. of } \{h_j; j \in S_n\} \\[2ex] \tilde{B}_n - \dfrac{(\tilde{B}_n h_{n+1})(\tilde{B}_n h_{n+1})^{\mathrm{T}}}{\sigma_{n+1}^2 + h_{n+1}^{\mathrm{T}} \tilde{B}_n h_{n+1}} \qquad \text{otherwise} \end{cases}$$

then for each n,

(b) $\tilde{x}_n$ is the vector of minimum norm among those which minimize $\sum_{j \in T_n} (\zeta_j - h_j^{\mathrm{T}} x)^2 / \sigma_j^2$ subject to $x \in \mathscr{C}_n$, provided $\mathscr{C}_n$ is nonempty. Furthermore, $\tilde{\varepsilon}_n$ is the associated residual error.

(8.5.6) **Example:** In (4.3) we showed how to relate $(H^{\mathrm{T}} \vdots h)^+$ to $H^{\mathrm{T}+}$. Using the results of (8.5), we can go in the opposite direction: Given $(H^{\mathrm{T}} \vdots h)^+$, we can compute $H^{\mathrm{T}+}$ without having to "start from scratch." The key to this procedure resides in the fact that the vector of minimum norm among those which minimize $\|y - Cv\|^2$ subject to the constraint $g^{\mathrm{T}} v = 0$ (where $C = (H^{\mathrm{T}} \vdots h)$ [an $m \times (n+1)$ matrix] and $g^{\mathrm{T}} = (0, 0, \ldots, 0, 1)$ [a $1 \times (n+1)$

row vector] is

$$\hat{\hat{v}} = \begin{pmatrix} H^{\mathrm{T}+} \\ \hdashline 0 \end{pmatrix} y. \tag{8.5.6.1}$$

This is a direct consequence of (3.12.4):

$$\hat{\hat{v}} = \bar{C}^{+}y \qquad \text{where} \quad \bar{C} = C(I-gg^{+}).$$

Since

$$I - gg^{+} = \begin{array}{c} \\ n \\ 1 \end{array}\!\!\begin{array}{c} \begin{array}{cc} n & 1 \end{array} \\ \begin{pmatrix} I & 0 \\ 0 & 0 \end{pmatrix} \end{array} \hat{\hat{v}}$$

$\bar{C} = (H^{\mathrm{T}} \vdots 0)$ and so

$$\bar{C}^{+} = \begin{pmatrix} H^{\mathrm{T}+} \\ \hdashline 0 \end{pmatrix}.$$

But the results of (8.5) tell us that the constraint

$$g^{\mathrm{T}}v = 0$$

can be thought of as a "new" observation $\zeta_{m+1} = 0$ with zero variance, and that

$$\hat{v} = C^{+}y$$

can be updated in the light of the new observation according to (8.3.1.1)–(8.3.1.4):

$$\hat{\hat{v}} = \hat{v} + K_C(\zeta_{m+1} - g^{\mathrm{T}}\hat{v}) = \hat{v} - K_C(g^{\mathrm{T}}\hat{v})$$

where

$$K_C = \begin{cases} \dfrac{(I-C^{+}C)g}{g^{\mathrm{T}}(I-C^{+}C)g} & \text{if} \quad g \text{ is not a l.c. of the rows of } C \\[2ex] \dfrac{(C^{\mathrm{T}}C)^{+}g}{g^{\mathrm{T}}(C^{\mathrm{T}}C)^{+}g} & \text{otherwise.} \end{cases}$$

But g is a l.c. of the rows of C if and only if g is a l.c. of the columns of C^{T} which is the same as $g \in \mathscr{R}(C^{\mathrm{T}})$ so that g is a l.c. of the rows of C if and only if

$$C^{+}Cg = g. \tag{3.7.10}$$

Since

$$C = (H^{\mathrm{T}} \vdots h) \qquad \text{and} \qquad g = (0, \ldots, 0, 1)$$

it follows that

$$Cg = h$$

so that

$$C^+Cg = C^+h.$$

Thus,

$$\begin{pmatrix} H^{\mathrm{T}+} \\ \hdashline 0 \end{pmatrix} y = \hat{\hat{v}} = (I - K_C g^{\mathrm{T}})\hat{v} = (I - K_C g^{\mathrm{T}})(H^{\mathrm{T}} \vdots h)^+ y \qquad \text{for all} \quad y$$

so that

(8.5.6.2)
$$\begin{pmatrix} H^{\mathrm{T}+} \\ \hdashline 0 \end{pmatrix} = (I - K_C g^{\mathrm{T}})(H^{\mathrm{T}} \vdots h)^+$$

where

$$K_C = \begin{cases} \dfrac{(I - C^+C)g}{g^{\mathrm{T}}(I - C^+C)g} & \text{if} \quad C^+h \neq g \\[2ex] \dfrac{(C^{\mathrm{T}}C)^+ g}{g^{\mathrm{T}}(C^{\mathrm{T}}C)^+ g} & \text{otherwise} \end{cases}$$

where $C = (H^{\mathrm{T}} \vdots h)$ and $g = (0, 0, \ldots, 0, 1)$. (8.5.6.2) is the desired result which relates $H^{\mathrm{T}+}$ to $(H^{\mathrm{T}} \vdots h)^+$.

(8.5.7) **Exercise** [*Continuation of* (8.5.6)]

(a) $C^+h = g$ if and only if $(I - H^+H)h \neq 0$.

(b) $$K_C = \begin{cases} (g - C^+h)/(1 - h^{\mathrm{T}}(CC^{\mathrm{T}})^+h) & \text{if} \quad (I - H^+H)h = 0 \\ C^+(CC^{\mathrm{T}})^+h/\|(CC^{\mathrm{T}})^+h\|^2 & \text{otherwise.} \end{cases}$$

(8.5.8) **Example** *Straight line regression performed recursively*

Suppose observations are collected which are of the form

(8.5.8.1)
$$\zeta_n = \alpha + \beta n + v_n \qquad n = 1, 2, 3$$

where the v's are uncorrelated, zero mean random variables with common variance. α and β are not known and are to be estimated.

Denote the least squares estimate (=BLUE) for $x = \binom{\alpha}{\beta}$, based on the first k observations by

$$\hat{x}_k = \begin{pmatrix} \hat{\alpha}_k \\ \hat{\beta}_k \end{pmatrix} \qquad k = 1, 2, 3.$$

If we let

$$h_k = \begin{pmatrix} 1 \\ k \end{pmatrix} \qquad k = 1, 2, 3$$

then $\hat{x}_k$ can be computed recursively via (8.3.1):

$$A_0 = I \qquad B_0 = 0 \qquad \hat{x}_0 = 0$$

$$K_1 = \frac{A_0 h_1}{h_1{}^{\mathrm{T}} A_0 h_1} = \begin{pmatrix} \frac{1}{2} \\ \frac{1}{2} \end{pmatrix}$$

$$\hat{x}_1 = K_1(\zeta_1 - h_1{}^{\mathrm{T}}\hat{x}_0) = \begin{pmatrix} \zeta_1/2 \\ \zeta_1/2 \end{pmatrix}.$$

$$A_1 = A_0 - h_1 h_1{}^{\mathrm{T}}/\|h_1\|^2$$

$$= \begin{pmatrix} 1 & 0 \\ 0 & 1 \end{pmatrix} - \begin{pmatrix} \frac{1}{2} & \frac{1}{2} \\ \frac{1}{2} & \frac{1}{2} \end{pmatrix} = \begin{pmatrix} \frac{1}{2} & -\frac{1}{2} \\ -\frac{1}{2} & \frac{1}{2} \end{pmatrix}$$

$$B_1 = (A_0 h_1)\frac{(A_0 h_1)^{\mathrm{T}}}{(h_1{}^{\mathrm{T}} A_0 h_1)^2} = \begin{pmatrix} \frac{1}{4} & \frac{1}{4} \\ \frac{1}{4} & \frac{1}{4} \end{pmatrix}$$

$$A_1 h_2 = \begin{pmatrix} -\frac{1}{2} \\ \frac{1}{2} \end{pmatrix} \qquad B_1 h_2 = \begin{pmatrix} \frac{3}{4} \\ \frac{3}{4} \end{pmatrix}.$$

Since h_2 is not a multiple of h_1,

$$K_2 = \frac{A_1 h_2}{h_2{}^{\mathrm{T}} A_1 h_2} = \begin{pmatrix} -1 \\ 1 \end{pmatrix}$$

$$\hat{x}_2 = \hat{x}_1 + K_2(\zeta_2 - h_2{}^{\mathrm{T}}\hat{x}_1)$$

$$= \begin{pmatrix} \zeta_1/2 \\ \zeta_1/2 \end{pmatrix} + \begin{pmatrix} -1 \\ 1 \end{pmatrix}(\zeta_2 - \tfrac{3}{2}\zeta_1)$$

$$= \begin{pmatrix} 2\zeta_1 - \zeta_2 \\ \zeta_2 - \zeta_1 \end{pmatrix}$$

$$A_2 = A_1 - (A_1 h_2)(A_1 h_2)^{\mathrm{T}}/(h_2{}^{\mathrm{T}} A_1 h_2) = 0$$

$$B_2 = B_1 - [(B_1 h_2)(A_1 h_2)^{\mathrm{T}} + (A_1 h_2)(B_1 h_2{}^{\mathrm{T}}]/h_2{}^{\mathrm{T}} A_1 h_2$$
$$+ (1 + h_2{}^{\mathrm{T}} B_1 h_2)(A_1 h_2)(A_1 h_2)^{\mathrm{T}}/(h_2{}^{\mathrm{T}} A_1 h_2)^2$$

$$= \begin{pmatrix} \frac{1}{4} & \frac{1}{4} \\ \frac{1}{4} & \frac{1}{4} \end{pmatrix} - \begin{pmatrix} -\frac{3}{2} & 0 \\ 0 & \frac{3}{2} \end{pmatrix} + \begin{pmatrix} \frac{13}{4} & -\frac{13}{4} \\ -\frac{13}{4} & \frac{13}{4} \end{pmatrix}$$

$$= \begin{pmatrix} 5 & -3 \\ -3 & 2 \end{pmatrix}.$$

$$B_2 h_3 = \begin{pmatrix} -4 \\ 3 \end{pmatrix}$$

and since h_3 is a linear combination of h_1 and h_2 ($h_3 = 2h_2 - h_1$),

$$K_3 = \frac{B_2 h_3}{(1 + h_3{}^{\mathrm{T}} B_2 h_3)} = \begin{pmatrix} -\frac{2}{3} \\ \frac{1}{2} \end{pmatrix}$$

(8.5.8.2)
$$\hat{x}_3 = \hat{x}_2 + K_3(\zeta_3 - h_3{}^{\mathrm{T}} \hat{x}_2)$$
$$= \begin{pmatrix} 4\zeta_1/3 + \zeta_2/3 - 2\zeta_3/3 \\ \zeta_3/2 - \zeta_1/2 \end{pmatrix}.$$

In Exercise (6.6.11), $\hat{\alpha}_3$ and $\hat{\beta}_3$ are given in closed form:

$$\hat{\beta}_3 = \left[\sum_{j=1}^{3} (\zeta_j - \bar{\zeta})(\tau_j - \bar{\tau})\right]\left[\sum_{j=1}^{3} (\tau_j - \bar{\tau})^2\right]^{-1}$$

and

$$\hat{\alpha}_3 = \bar{\zeta} - \hat{\beta}_3 \bar{\tau}$$

where

$$\bar{\zeta} = \tfrac{1}{3} \sum_{j=1}^{3} \zeta_j \qquad \text{and} \qquad \bar{\tau} = \tfrac{1}{3} \sum_{j=1}^{3} \tau_j .$$

In the present case, $\tau_j = j$ so $\bar{\tau} = 2$. Thus

$$\hat{\beta}_3 = \frac{-(\zeta_1 - \bar{\zeta}) + (\zeta_3 - \bar{\zeta})}{2} = \frac{\zeta_3 - \zeta_1}{2}$$

and

$$\hat{\alpha}_3 = \bar{\zeta} - 2\left(\frac{\zeta_3 - \zeta_1}{2}\right) = \frac{\zeta_1 + \zeta_2 + \zeta_3}{3} - \zeta_3 + \zeta_1$$
$$= \frac{4\zeta_1}{3} + \frac{\zeta_2}{3} - \frac{2\zeta_3}{3}$$

which agrees with (8.5.8.2).

The covariance for $\hat{x}_3$ is given by $\sigma^2 B_3$ where σ^2 is the variance of the v's and

(8.5.8.3)
$$B_3 = B_2 - (B_2 h_3)(B_2 h_3)^{\mathrm{T}}/(1 + h_3{}^{\mathrm{T}} B_2 h_3)$$
$$= \begin{pmatrix} \frac{7}{3} & -1 \\ -1 & \frac{1}{2} \end{pmatrix}.$$

($A_3 = 0$ since h_1 and h_2 span two-dimensional space.)

(8.5.9) **Adding a Constraint**

Suppose we now wish to compute the least squares estimate for $x = \binom{\alpha}{\beta}$ based on the first three observations, subject to the constraint $\beta = 0$.

According to (8.5.1), this may be achieved by adding the fictitious observation $\zeta_4 = 0$ "pretending" at the same time that

$$\zeta_4 = h_4^{\mathrm{T}} x + v_4$$

where v_4 has zero variance and

$$h_4 = \begin{pmatrix} 0 \\ 1 \end{pmatrix}.$$

We apply (8.3.1) with $\sigma_4^2 = 0$ and find that

$$K_4 = \frac{B_3 h_4}{h_4^{\mathrm{T}} B_3 h_4} = \begin{pmatrix} -2 \\ 1 \end{pmatrix}$$

(8.5.9.1)
$$\hat{x}_4 = \hat{x}_3 + K_4(0 - h_4^{\mathrm{T}} \hat{x}_3)$$

$$= (I - K_4 h_4^{\mathrm{T}}) \hat{x}_3 = \begin{pmatrix} \frac{1}{3}(\zeta_1 + \zeta_2 + \zeta_3) \\ 0 \end{pmatrix}$$

$$= \begin{pmatrix} \bar{\zeta} \\ 0 \end{pmatrix}$$

which is what we should get if there is any justice.

(8.5.10) **Incorporating the Constraint**

Using (8.4), we could carry the constraint right along and compute the least squares estimate for x based on one, two, and three observations, all subject to $\beta = 0$.

The constraint $\beta = 0$ is the same as $Gx = 0$ where $G = (0 \quad 1)$. Then

$$G^+ = \begin{pmatrix} 0 \\ 1 \end{pmatrix} \qquad I - G^+ G = \begin{pmatrix} 1 & 0 \\ 0 & 0 \end{pmatrix}$$

$$\bar{h}_k = (I - G^+ G) h_k = \begin{pmatrix} 1 \\ 0 \end{pmatrix} \qquad k = 1, 2, 3.$$

Therefore, $\bar{h}_2$ is a l.c. of $\bar{h}_1$ as is $\bar{h}_3$ and

$$\bar{K}_1 = \frac{\bar{A}_0 \bar{h}_1}{\bar{h}_1^{\mathrm{T}} \bar{A}_0 \bar{h}_1} = \begin{pmatrix} 1 \\ 0 \end{pmatrix}$$

$$\bar{x}_1 = \bar{K}_1(\zeta_1) = \begin{pmatrix} \zeta_1 \\ 0 \end{pmatrix}$$

$$\bar{A}_1 = \frac{\bar{A}_0 - \bar{h}_1 \bar{h}_1^{\mathrm{T}}}{\|\bar{h}_1\|^2} = \begin{pmatrix} 0 & 0 \\ 0 & 1 \end{pmatrix}$$

$$\bar{B}_1 = \frac{(\bar{A}_0 \bar{h}_1)(\bar{A}_0 \bar{h}_1)^{\mathrm{T}}}{(\bar{h}_1^{\mathrm{T}} \bar{A}_0 \bar{h}_1)^2} = \begin{pmatrix} 1 & 0 \\ 0 & 0 \end{pmatrix}$$

$$\bar{K}_2 = \frac{\bar{B}_1 \bar{h}_2}{(1 + \bar{h}_2^{\mathrm{T}} \bar{B}_1 \bar{h}_2)} = \begin{pmatrix} \frac{1}{2} \\ 0 \end{pmatrix}$$

$$\bar{x}_2 = \bar{x}_1 + \bar{K}_2(\zeta_2 - \bar{h}_2^{\mathrm{T}} \bar{x}_1) = \begin{pmatrix} \frac{1}{2}(\zeta_1 + \zeta_2) \\ 0 \end{pmatrix}$$

$$\bar{A}_2 = \bar{A}_1$$

$$\bar{B}_2 = \bar{B}_1 - \frac{(\bar{B}_1 \bar{h}_1)(\bar{B}_1 \bar{h}_1)^{\mathrm{T}}}{(1 + \bar{h}_1^{\mathrm{T}} \bar{B}_1 \bar{h}_1)} = \begin{pmatrix} \frac{1}{2} & 0 \\ 0 & 0 \end{pmatrix}$$

$$K_3 = \frac{\bar{B}_2 \bar{h}_3}{(1 + \bar{h}_3^{\mathrm{T}} \bar{B}_2 \bar{h}_3)} = \begin{pmatrix} \frac{1}{3} \\ 0 \end{pmatrix}$$

$$\bar{x}_3 = \bar{x}_2 + \bar{K}_3(\zeta_3 - \bar{h}_3^{\mathrm{T}} \bar{x}_2) = \begin{pmatrix} \frac{1}{3}(\zeta_1 + \zeta_2 + \zeta_3) \\ 0 \end{pmatrix} = \begin{pmatrix} \bar{\zeta} \\ 0 \end{pmatrix}$$

which agrees with (8.5.9.1).

(8.6) Additional Regressors, II (Stepwise Regression)

In (4.4) we discussed the problem of enriching a linear model by adding additional regressors and we derived a recursive relationship between the best m-dimensional regression and the best $(m+1)$-dimensional regression in (4.4.4). That result is limited to the situation where regressors are added one at a time. This restriction is not essential. Indeed, the results of (6.6.14) can be applied to yield immediate results for the general case:

Suppose then, that a "preliminary" model of the form

(8.6.1) $$\mathbf{z} = H_1 x_1 + \mathbf{v}$$

is analyzed, where H_1 is an $n \times p$ matrix, and $\mathbf{v}$'s covariance is proportional to the identity matrix. The BLUE for x_1 would then be

(8.6.2) $$\mathbf{x}_1{}^* = H_1{}^+ \mathbf{z}$$

with the associated residual error

(8.6.3) $$\boldsymbol{\varepsilon}_1 = \|\mathbf{z} - H_1 \mathbf{x}_1{}^*\|^2.$$

If $\boldsymbol{\varepsilon}_1$ were judged to be too large, an enriched model of the form

(8.6.4) $$\mathbf{z} = (H_1 \vdots H_2)\begin{pmatrix} x_1 \\ \hdashline x_2 \end{pmatrix} + \mathbf{v}$$

might be investigated, where H_2 is some suitably chosen $n \times q$ matrix.

The BLUE for

$$\begin{pmatrix} x_1 \\ \hdashline x_2 \end{pmatrix}$$

would now be

(8.6.5) $$\begin{pmatrix} \hat{\mathbf{x}}_1 \\ \hdashline \hat{\mathbf{x}}_2 \end{pmatrix} = (H_1 \vdots H_2)^+ \mathbf{z}$$

and the residual error would be

(8.6.6) $$\boldsymbol{\varepsilon}_{12} = \|\mathbf{z} - H_1 \hat{\mathbf{x}}_1 - H_2 \hat{\mathbf{x}}_2\|^2.$$

If $(H_1 \vdots H_2)$ has rank $p+q$, we can apply the results of (6.6.14) to these computations: Letting $Q_1 = I - H_1 H_1{}^+$ we find that

(8.6.7) $$\hat{\mathbf{x}}_2 = (Q_1 H_2)^+ \mathbf{z}$$

(8.6.8) $$\hat{\mathbf{x}}_1 = H_1{}^+(\mathbf{z} - H_2 \hat{\mathbf{x}}_2)$$

(8.6.9) $$\boldsymbol{\varepsilon}_{12} = \boldsymbol{\varepsilon}_1 - \|Q_1 H_2 \hat{\mathbf{x}}_2\|^2.$$

(8.6.10) **Exercise**

(a) If $(H_1 \vdots H_2)$ has full column rank then $H_2{}^{\mathrm{T}} Q_1 H_2$ is nonsingular.
(b) In this case

$$H_2{}^{\mathrm{T}} Q_1 H_2 \hat{\mathbf{x}}_2 = H_2{}^{\mathrm{T}} Q_1 \mathbf{z}$$

and

(c) $\boldsymbol{\varepsilon}_{12} = \boldsymbol{\varepsilon}_1 - (H_2 \hat{\mathbf{x}}_2)^{\mathrm{T}}(Q_1 \mathbf{z})$.

(8.7) Example *Relationship between analysis of variance and analysis of covariance* (Scheffé [1, Section 6.3]): Let

(8.7.1)
$$\mathbf{z} = \begin{bmatrix} \zeta_{11} \\ \zeta_{12} \\ \zeta_{1J_1} \\ \zeta_{21} \\ \vdots \\ \zeta_{kJ_k} \end{bmatrix}$$

and suppose a model for the observations of the form

(8.7.2)
$$\zeta_{ij} = \beta_i + \mathbf{v}_{ij} \qquad j = 1,\ldots,J_i; \quad i = 1,\ldots,k,$$

were assumed (one-way layout), where the J_{ij}'s were uncorrelated zero mean random variables with common variance.

The BLUE (=least squares estimate) for the vector of β_i's is unique and easily seen to be given by

(8.7.3)
$$\boldsymbol{\beta}_i{}^* = (J_i)^{-1} \sum_j \zeta_{ij} \equiv \zeta_i.$$

which may be verified by differentiating

$$\sum_{ij} (\zeta_{ij} - \beta_i)^2$$

with respect to the β_i's, setting the results equal to zero and solving the resulting normal equations.

The associated residual sum of squares is

(8.7.4)
$$\boldsymbol{\varepsilon}_1 = \sum_{ij} (\zeta_{ij} - \zeta_i.)^2.$$

In vector notation, we write

$$\mathbf{z} = H_1 x_1 + \mathbf{v}$$

where

$$x_1 = \begin{bmatrix} \beta_1 \\ \vdots \\ \beta_k \end{bmatrix}.$$

The BLUE for x_1 is

(8.7.5)
$$\mathbf{x}_1{}^* = \begin{pmatrix} \boldsymbol{\beta}_1{}^* \\ \vdots \\ \boldsymbol{\beta}_k{}^* \end{pmatrix} = H_1{}^+ \mathbf{z} = \begin{pmatrix} \zeta_1. \\ \vdots \\ \zeta_k. \end{pmatrix}$$

and

$$\varepsilon_1 = \|\mathbf{z}-H_1\mathbf{x}_1{}^*\|^2 = \|(I-H_1H_1{}^+)\mathbf{z}\|^2 = \|Q_1\mathbf{z}\|^2 = \sum_{ij}(\zeta_{ij}-\zeta_{i\cdot})^2. \tag{8.7.6}$$

Equation (8.7.5) tells us that the operator $H_1{}^+$ transforms any vector z which is written in the double subscript form (8.7.1), into the vector $H_1{}^+z$ whose components are obtained by averaging the components of z having the same initial subscript.

Symbolically,

$$H_1{}^+\begin{bmatrix}\zeta_{11}\\ \zeta_{12}\\ \vdots\\ \zeta_{kJ_k}\end{bmatrix} = \begin{bmatrix}\zeta_{1\cdot}\\ \vdots\\ \zeta_{k\cdot}\end{bmatrix}. \tag{8.7.7}$$

By the same token, (8.7.6) tells us that Q_1 transforms any vector z of the form (8.7.1) into a vector

$$\bar{z} = Q_1 z = \begin{bmatrix}\bar{\zeta}_{11}\\ \bar{\zeta}_{12}\\ \vdots\\ \bar{\zeta}_{kJ_k}\end{bmatrix} \tag{8.7.8}$$

where

$$\bar{\zeta}_{ij} = \zeta_{ij} - \zeta_{i\cdot}. \tag{8.7.9}$$

If ε_1 is judged to be too large and the model is enriched by the addition of a regression term of the form $\gamma\eta_{ij}$ (where γ is unknown and the η_{ij}'s are known), the model takes the form

$$\zeta_{ij} = \beta_i + \gamma\eta_{ij} + \mathbf{v}_{ij}. \tag{8.7.10}$$

In the vector notation

$$\mathbf{z} = H_1x_1 + H_2x_2 + \mathbf{v} \tag{8.7.11}$$

where H_2 is a matrix with one column (i.e., a vector) and x_2 has one component:

$$x_2 = [\gamma] \tag{8.7.12}$$

$$H_2 = \begin{bmatrix}\eta_{11}\\ \eta_{12}\\ \vdots\\ \eta_{kJ_k}\end{bmatrix}. \tag{8.7.13}$$

Applying (8.6.7), we see that the BLUE for x_2 is

(8.7.14) $$\hat{\mathbf{x}}_2 = (Q_1 H_2)^+ \mathbf{z}.$$

By virtue of (8.7.8),

(8.7.15) $$Q_1 H_2 = \bar{H}_2 = \begin{bmatrix} \bar{\eta}_{11} \\ \vdots \\ \bar{\eta}_{kJ_k} \end{bmatrix}$$

where

(8.7.16) $$\bar{\eta}_{ij} = \eta_{ij} - \eta_{i\cdot}$$

We know that for any matrix with one column,

(8.7.17) $$\bar{H}_2^{\,+} = (\bar{H}_2^{\,\mathrm{T}} \bar{H}_2)^{-1} \bar{H}_2^{\,\mathrm{T}} = \bar{H}_2^{\,\mathrm{T}} / \|\bar{H}_2\|^2,$$

and so

$$\hat{\mathbf{x}}_2 = (\bar{H}_2^{\,\mathrm{T}} \mathbf{z}) / \|\bar{H}_2\|^2$$

or equivalently

(8.7.18) $$\hat{\gamma} = \sum_{ij} \bar{\eta}_{ij} \zeta_{ij} \Big/ \sum_{ij} \bar{\eta}_{ij}^2 .$$

The BLUE for the β's in the enriched model are found using (8.6.8):

(8.7.19) $$\hat{\mathbf{x}}_1 = H_1^{\,+} (\mathbf{z} - H_2 \hat{\mathbf{x}}_2).$$

But

(8.7.20) $$\mathbf{z} - H_2 \hat{\mathbf{x}}_2 \equiv \mathbf{z}^* = \begin{bmatrix} \zeta_{11}^* \\ \vdots \\ \zeta_{kJ_k}^* \end{bmatrix}$$

where

(8.7.21) $$\zeta_{ij}^* = \zeta_{ij} - \hat{\gamma} \eta_{ij} .$$

By virtue of (8.7.7), the ith component of $H_1^{\,+} \mathbf{z}^*$ is $\zeta_{i\cdot}^*$, so

(8.7.22) $$\hat{\mathbf{x}}_1 = \begin{pmatrix} \hat{\boldsymbol{\beta}}_1 \\ \vdots \\ \hat{\boldsymbol{\beta}}_k \end{pmatrix} = \begin{pmatrix} \zeta_{1\cdot}^* \\ \vdots \\ \zeta_{k\cdot}^* \end{pmatrix}.$$

Finally, the residual sum of squares $\boldsymbol{\varepsilon}_{12}$, is obtained via (8.6.9):

$$\boldsymbol{\varepsilon}_{12} \equiv \|\mathbf{z} - H_1 \hat{\mathbf{x}}_1 - H_2 \hat{\mathbf{x}}_2\|^2 = \boldsymbol{\varepsilon}_1 - \hat{\gamma}^2 \|Q_1 H_2\|^2 .$$

By virtue of (8.7.15), the last is equal to

(8.7.23) $$\boldsymbol{\varepsilon}_{12} = \boldsymbol{\varepsilon}_1 - \hat{\gamma}^2 \sum_{ij} \bar{\eta}_{ij}^2 = \sum_{ij} (\bar{\zeta}_{ij}^2 - \hat{\gamma}^2 \bar{\eta}_{ij}^2).$$

(8.8) Missing Observations

In (6.6) we discussed the computational advantages of orthogonal designs when computations are carried out on a desk calculator. But the best laid plans of men for mice often go haywire and it is painfully common for something to foul up the experimental design during the course of the experiment (e.g., an experimental animal dies) thereby causing an experimental observation to be ruined, and destroying the orthogonality of the design. For example, suppose the experiment were planned in such a way that the observation vector z is of the form

$$z = Hx + v$$

where x is the (k-dimensional vector) parameter of interest. Suppose further, that the experiment had been carefully designed so that the columns of H would be mutually orthogonal and have unit lengths. Then the least squares estimate would be marvelously easy to compute:

$$\hat{x} = H^{+}z = (H^{T}H)^{-1}H^{T}z = H^{T}z$$

(since $H^{T}H = I$ if the columns of H are orthonormal).

If a rat dies before the experiment is over, an observation is ruined. The actual vector of observation is now

$$z^{*} = H^{*}x + v$$

where H^{*} is a maimed version of H: One row is missing so that H^{*}'s columns are no longer orthogonal. It looks as though we are stuck with the job of psuedoinverting H^{*} or something equally messy, right?

Wrong! For the sake of concreteness, assume that the missing observation corresponds to the last one, so that if H is an $n \times k$ matrix, H^{*} is an $(n-1) \times k$ matrix and

$$H = \begin{pmatrix} H^{*} \\ \hdashline h^{T} \end{pmatrix}$$

(where h^{T} is the last row vector of H, the one that is missing from H^{*}).

By (8.5.6.2)

$$[(H^{*})^{+} \,\vdots\, 0] = H^{T}(I - gK^{T})$$

where

$$g = \begin{pmatrix} 0 \\ 0 \\ \vdots \\ 1 \end{pmatrix} \qquad (n \text{ components})$$

and

$$K = \begin{cases} \dfrac{(I-HH^+)g}{g^{\mathrm{T}}(I-HH^+)g} & \text{if } g \text{ is not a l.c. of the columns of } H \\[2ex] \dfrac{(HH^{\mathrm{T}})^+ g}{g^{\mathrm{T}}(HH^{\mathrm{T}})^+ g} & \text{otherwise.} \end{cases}$$

Since the columns of H (call them $g_1, g_2, \ldots, g_k$) are orthonormal, the question of whether g is a l.c. of $g_1, \ldots, g_k$ is easily settled:

$$HH^+ g = HH^{\mathrm{T}} g = \sum_{j=1}^{k} (g_j^{\mathrm{T}} g) g_j$$

so that one need only compare g to the right side of the last equation to see whether or not $(I-HH^+)g = 0$. Notice also that $(HH^{\mathrm{T}})^+ = HH^{\mathrm{T}}$, (3.7.3).

Thus

$$[(H^*)^+ : 0] = H^{\mathrm{T}}(I - gK^{\mathrm{T}})$$

where

$$K = \begin{cases} (I-HH^{\mathrm{T}})g/(1-\|H^{\mathrm{T}}g\|^2) & \text{if} \quad HH^{\mathrm{T}}g \neq g \\ HH^{\mathrm{T}}g/\|HH^{\mathrm{T}}g\|^2 & \text{if} \quad HH^{\mathrm{T}}g = g \end{cases}$$

$$= \begin{cases} (I-HH^{\mathrm{T}})g/(1-\|H^{\mathrm{T}}g\|^2) & \text{if} \quad HH^{\mathrm{T}}g \neq g \\ g & \text{otherwise.} \end{cases}$$

In this case, a very simple relationship exists between the pseudoinverse of the matrix H (whose columns are orthonormal) and the pseudoinverse of H^*, the matrix obtained by deleting H's last row.

The least squares estimate for x and the corresponding residual sum of squares for the "censored" data, z^*, is thereby simply related to that which would have been obtained if the rat had not died:

$$\hat{x}^* = H^{*+} z^* = [(H^*)^+ \vdots 0] z$$

(since

$$z = \begin{pmatrix} z^* \\ \hdashline \zeta \end{pmatrix}$$

where ζ is the missing observation) so

$$\hat{x}^* = H^{\mathrm{T}}(I - gK^{\mathrm{T}}) z = H^{\mathrm{T}} \bar{z}$$

where

$$\bar{z} = z - g(K^{\mathrm{T}} z)$$

is a modified observation vector which agrees with the uncensored sample z, in the first n components, but "pretends" that the missing observation was actually observed and had the value $\zeta - K^{\mathrm{T}}z$.

Since $K^{\mathrm{T}}g = 1$, it follows that

$$(I - gK^{\mathrm{T}})g = g(I - K^{\mathrm{T}}g) = 0$$

so that

$$(I - gK^{\mathrm{T}})(z_1 - z_2) = 0$$

if z_1 and z_2 agree in their first $n-1$ components. Therefore, $\bar{z}$ does not depend explicitly on the choice of z's last component, ζ, and we can choose $\zeta = 0$.

This analysis can be generalized to the case of a missing observation for an arbitrary design.

(8.8.1) **Theorem:** Let

$$H = \begin{pmatrix} \tilde{H} \\ \hline h^{\mathrm{T}} \end{pmatrix} \qquad z = \begin{pmatrix} \tilde{z} \\ \hline 0 \end{pmatrix} \qquad B = (H^{\mathrm{T}}H)^{+}$$

$$\hat{x} = H^{+}z \qquad \varepsilon = \|z - H\hat{x}\|^2 \qquad \tilde{x} = \tilde{H}^{+}\tilde{z} \qquad \text{and} \qquad \tilde{\varepsilon} = \|\tilde{z} - \tilde{H}\tilde{x}\|^2.$$

Then

(8.8.1.1)
$$\tilde{x} = (I - Bhh^{\mathrm{T}}M)\hat{x}$$

where

$$M = \begin{cases} -I/(1 - h^{\mathrm{T}}Bh) & \text{if } h \text{ is a l.c. of the rows of } \tilde{H} \\ B/\|Bh\|^2 & \text{otherwise} \end{cases}$$

(8.8.1.2)
$$\tilde{\varepsilon} = \begin{cases} \varepsilon - (h^{\mathrm{T}}\hat{x})^2/(1 - h^{\mathrm{T}}Bh) & \text{if } h \text{ is a l.c. of the rows of } \tilde{H} \\ \varepsilon & \text{otherwise} \end{cases}$$

Comment: $\hat{x}$ is the least squares estimate for x based on the full data model

$$z = Hx + v,$$

and ε is the associated residual sum of squares.

If an observation is missing (say the last), then the observations are of the form

$$\tilde{z} = \tilde{H}x + \tilde{v}$$

where $\tilde{H}$ is obtained from H by deleting its last row. The new least squares estimate is

$$\tilde{x} = \tilde{H}^{+}\tilde{z}$$

and its residual sum of squares is $\tilde{\varepsilon} = \|\tilde{z} - \tilde{H}\tilde{x}\|^2$.

If more than one observation is missing, the recursion (8.8.1) can be iterated as many times as necessary. However, care must be exercised in defining $\tilde{B} = (\tilde{H}^T\tilde{H})^+$ for all cases [cf. (8.8.2)], not just in the case where h is a l.c. of $\tilde{H}$'s rows.

Proof of theorem: $\tilde{x} = (\tilde{H}^+ \vdots 0)z = H^+(I - gk^T)z$ where

$$k = \begin{cases} (g - H^{T+}h)/1 - h^T(H^TH)^+h & \text{if } h \text{ is a l.c. of the rows of } \tilde{H} \\ H^{T+}(H^TH)^+h/\|(H^TH)^+h\|^2 & \text{otherwise.} \end{cases} \qquad \text{(8.5.6.2) and (8.5.7)}$$

Since z's last component is zero, $g^Tz = 0$ so that

$$k^Tz = \begin{cases} -h^T\hat{x}/1 - h^TBh & \text{if } h \text{ is a l.c. of } \tilde{H}\text{'s rows} \\ h^TB\hat{x}/\|Bh\|^2 & \text{otherwise.} \end{cases}$$

(8.8.1.1) follows once we observe that

$$H^+g = BH^Tg = B(\tilde{H}^T \vdots h)g = Bh$$

since g has all zero components except the last.

The formula for $\tilde{\varepsilon}$ follows from (8.2.2).

(8.8.2) **Exercise:** Let $\tilde{B} = (\tilde{H}^T\tilde{H})^+$. Then

$$\text{(a)} \qquad \tilde{B} = \begin{cases} B - \dfrac{(Bh)(B^2h)^T + (B^2h)(Bh)^T}{\|Bh\|^2} & \\ \quad + \dfrac{h^TB^3h}{\|Bh\|^4}(Bh)(Bh)^T & \text{if } h \text{ is not a l.c. of } \tilde{H}\text{'s rows} \\ B + \dfrac{(Bh)(Bh)^T}{1 - h^TBh} & \text{otherwise.} \end{cases}$$

(b) $\quad h^TBh < 1 \quad$ if h is a l.c. of $\tilde{H}$'s rows.

(8.8.3) **Example:** Suppose

$$H = \begin{pmatrix} 1 \\ 1 \\ \vdots \\ 1 \end{pmatrix} (n \times 1) \qquad \text{and} \qquad \tilde{H} = \begin{pmatrix} 1 \\ 1 \\ \vdots \\ 1 \end{pmatrix} (n-1) \times 1.$$

Let

$$\tilde{z} = \begin{pmatrix} \zeta_1 \\ \vdots \\ \zeta_{n-1} \end{pmatrix} \qquad \text{and} \qquad z = \begin{pmatrix} \tilde{z} \\ \hline 0 \end{pmatrix}.$$

Then $h = (1)$, $B = (H^{\mathrm{T}}H)^{-1} = n^{-1}$,

$$\hat{x} = H^{+}z = n^{-1}\sum_{j=1}^{n}\zeta_j = n^{-1}\sum_{j=1}^{n-1}\zeta_j$$

$$\varepsilon = \sum_{j=1}^{n}(\zeta_j - \hat{x})^2 = \sum_{j=1}^{n-1}\zeta_j^2 - n\hat{x}^2.$$

From (8.8.1)

$$M = -1/(1-n^{-1}) = -n/(n-1)$$

$$\tilde{x} = \left[1 - (n^{-1})\left(\frac{-n}{n-1}\right)\right]\hat{x} = \left(1 + \frac{1}{n-1}\right)\hat{x}$$

$$= \left(\frac{n}{n-1}\right)\left(n^{-1}\sum_{j=1}^{n-1}\zeta_j\right) = \frac{1}{n-1}\sum_{j=1}^{n-1}\zeta_j$$

and

$$\tilde{\varepsilon} = \varepsilon - \hat{x}^2 \bigg/ \left(1 - \frac{1}{n}\right)$$

$$= \sum_{j=1}^{n-1}\zeta_j^2 - \frac{n^2\hat{x}^2}{n-1}.$$

But $\tilde{x} = [n/(n-1)]\,\hat{x}$, so

$$\frac{n^2\hat{x}^2}{n-1} = (n-1)\,\tilde{x}^2$$

hence

$$\tilde{\varepsilon} = \sum_{j=1}^{n-1}\zeta_j^2 - (n-1)\,\tilde{x}^2$$

as it should.

Chapter IX

NONNEGATIVE DEFINITE MATRICES, CONDITIONAL EXPECTATION, AND KALMAN FILTERING

(9.1) Nonnegative Definiteness

It is well known that the following conditions are equivalent definitions for nonnegative definiteness of a symmetric matrix S, (Bellman [1]):

(9.1.1) $S = HH^{\mathrm{T}}$ for some matrix H;

(9.1.2) $x^{\mathrm{T}}Sx \geqslant 0$ for all vectors x;

(9.1.3) The eigenvalues of S are nonnegative;

(9.1.4) There is a symmetric matrix R such that $R^2 = S$. (R is called the square root of S, and is denoted by the symbol, $S^{1/2}$.)

If S is nonnegative definite *and* nonsingular, it is said to be *positive definite*. In this case, $S^{1/2}$ is also positive definite and the inequality in (9.1.2) is strong for all nonzero x.

The statement "S is nonnegative definite" is abbreviated "$S \geqslant 0$." Similarly, "$S > 0$" means "S is positive definite."

(9.1.5) **Exercise:** If $S \geqslant 0$ and $T \geqslant 0$ then $S+T \geqslant 0$ with strict inequality holding if and only if $\mathcal{N}(S) \cap \mathcal{N}(T) = \{0\}$.

One of the classical results concerning nonnegative definiteness, states that

S is nonnegative definite (positive definite) if and only if the principal subdeterminants of S are all nonnegative (positive), (Bellman [1]). This result reduces the decision about S's nonnegativeness to the evaluation of a number of determinants.

The material which we have developed in earlier chapters concerning pseudoinverses, can be used to prove some new results about nonnegative (positive) definiteness.

(9.1.6) **Theorem** (Albert [1]): Let S be a square matrix:

$$S = \left(\begin{array}{c|c} S_{11} & S_{12} \\ \hline S_{12}^{\mathrm{T}} & S_{22} \end{array}\right)$$

where S_{11} is a symmetric $n \times n$ matrix and S_{22} is a symmetric $m \times m$ matrix. Then

(a) $S \geqslant 0$ if and only if $S_{11} \geqslant 0$, $S_{11} S_{11}^{+} S_{12} = S_{12}$ and

$$S_{22} - S_{12}^{\mathrm{T}} S_{11}^{+} S_{12} \geqslant 0.$$

(b) $S > 0$ if and only if $S_{11} > 0$, $S_{22} > 0$, $S_{11} - S_{12} S_{22}^{-1} S_{12}^{\mathrm{T}} > 0$ and

$$S_{22} - S_{12}^{\mathrm{T}} S_{11}^{-1} S_{12} > 0.$$

Proof: (a) (Sufficiency): If $S \geqslant 0$ then $S = HH^{\mathrm{T}}$, (9.1.1), where H has $n+m$ rows. Write H as a partitioned matrix:

$$H = \begin{matrix} n \\ m \end{matrix}\left(\begin{array}{c} X \\ \hline Y \end{array}\right).$$

Then,

$$S = HH^{\mathrm{T}} = \left(\begin{array}{c|c} XX^{\mathrm{T}} & XY^{\mathrm{T}} \\ \hline YX^{\mathrm{T}} & YY^{\mathrm{T}} \end{array}\right),$$

so that

$$S_{11} = XX^{\mathrm{T}} \geqslant 0 \qquad \text{and} \qquad S_{12} = XY^{\mathrm{T}}.$$

By (3.11.9),

$$S_{11} S_{11}^{+} = (XX^{\mathrm{T}})(XX^{\mathrm{T}})^{+} = XX^{+}$$

so that

$$(S_{11} S_{11}^{+}) S_{12} = (XX^{+}) XY^{\mathrm{T}} = XY^{\mathrm{T}} = S_{12}.$$

Finally, if we let

$$U = Y - S_{12}^{\mathrm{T}} S_{11}^{+} X$$

then

$$0 \leqslant UU^{\mathrm{T}} = S_{22} - S_{12}^{\mathrm{T}} S_{11}^{+} S_{12}.$$

(Necessity): Let

$$U = {}_n(\overset{n}{I}\,\vdots\,\overset{m}{0}) \qquad V = {}_m(\overset{n}{0}\,\vdots\,\overset{m}{I}) \qquad X = S_{11}^{1/2}\,U$$

$$Y = S_{12}^{\mathrm{T}}\,S_{11}^{+}\,S_{11}^{1/2}\,U + (S_{22} - S_{12}^{\mathrm{T}}\,S_{11}^{+}\,S_{12})^{1/2}V.$$

Since $UV^{\mathrm{T}} = {}_n\overset{m}{0}$, we see that

$$0 \leqslant \begin{pmatrix} X \\ \hline Y \end{pmatrix}\begin{pmatrix} X \\ \hline Y \end{pmatrix}^{\mathrm{T}} = \left(\begin{array}{c|c} S_{11} & S_{12} \\ \hline S_{12}^{\mathrm{T}} & S_{22} \end{array}\right) = S.$$

(b) (Sufficiency): If $S > 0$ then by part (a), $S_{11} \geqslant 0$. Furthermore, $\det S_{11} > 0$ since all principal subdeterminants of S are positive. Hence S_{11} is nonsingular and therefore $S_{11} > 0$. Similarly, $S_{22} - S_{12}^{\mathrm{T}}\,S_{11}^{-1}\,S_{12} \geqslant 0$ so that $S_{22} \geqslant 0$ and by the same argument $S_{22} > 0$.

The eigenvalues of S^{-1} are the reciprocals of S's and so, $S^{-1} > 0$ if $S > 0$, (9.1.3). We can therefore write

$$S^{-1} = \begin{array}{c} \\ n \\ m \end{array}\!\!\left(\begin{array}{c|c} \overset{n}{A} & \overset{m}{B} \\ \hline B^{\mathrm{T}} & C \end{array}\right) > 0$$

where the condition

$$SS^{-1} = \left(\begin{array}{c|c} I & 0 \\ \hline 0 & I \end{array}\right)$$

dictates that

$$(S_{11} - S_{12}\,S_{22}^{-1}\,S_{12}^{\mathrm{T}})A = I$$

and

$$(S_{22} - S_{12}^{\mathrm{T}}\,S_{11}^{-1}\,S_{12})\,C = I.$$

$A > 0$ and $C > 0$ since $S^{-1}0$. Hence $A^{-1} > 0$ and $C^{-1} > 0$ by the same argument as above. This proves sufficiency.

(Necessity): From part (a), $S \geqslant 0$. Let

$$A = (S_{11} - S_{12}\,S_{22}^{-1}\,S_{12}^{\mathrm{T}})^{-1}$$

$$B = -S_{11}^{-1}\,S_{12}(S_{22} - S_{12}^{\mathrm{T}}\,S_{11}^{-1}\,S_{12})^{-1}$$

and

$$C = (S_{22} - S_{12}^{\mathrm{T}}\,S_{11}^{-1}S_{12})^{-1}.$$

It is easy to show that

$$A^{-1}BC^{-1} = A^{-1}[-(S_{11} - S_{12}\,S_{22}^{-1}S_{12})^{-1}\,S_{12}\,S_{22}^{-1}]\,C^{-1}$$

so that

$$B = -(S_{11} - S_{12} S_{22}^{-1} S_{12})^{-1} S_{12} S_{22}^{-1}.$$

Routine calculations then verify that

$$S \begin{pmatrix} A & B \\ B^{\mathrm{T}} & C \end{pmatrix} = I$$

so S is nonsingular. ■

Comment: In earlier chapters we pointed out that the class of covariance matrices coincides with the class of nonnegative-definite matrices. The preceding theorem tells us that the partitioned submatrices of covariance matrices have particular properties. This has a statistical interpretation which we will exploit in the section on conditional expectations.

(9.1.7) **Exercise:** Let S_N be an $N \times N$ symmetric matrix whose (i,j)th element is $\sigma(i,j)$. Let S_k be the $k \times k$ submatrix occupying the upper left corner of S_N, and let s_k be the k-dimensional column vector (just to the right of S_k) with components $\sigma(k+1,1)$, $\sigma(k+1,2), \ldots, \sigma(k+1,k)$. Define $S_0 = 0$ and $s_0 = 0$. Then

(a) $S_N \geqslant 0$ if and only if $S_{k-1} S_{k-1}^{+} s_{k-1} = s_{k-1}$

and

$$\sigma(k,k) \geqslant s_{k-1}^{\mathrm{T}} S_{k-1}^{+} s_{k-1} \qquad \text{for} \qquad k = 1,2,\ldots,N.$$

(b) $S_N > 0$ if and only if $\sigma(k,k) > s_{k-1}^{\mathrm{T}} S_{k-1}^{+} s_{k-1}$ for $k = 1,2,\ldots,N$. [*Hint:* Use (9.1.6) and proceed by induction. Notice that if $S_k > 0$, then $[S_k^{1/2}(I - xx^{\mathrm{T}}) S_k^{1/2}] > 0$ if and only if $I - xx^{\mathrm{T}} > 0$ which is true if and only if $x^{\mathrm{T}} x < 1$. Apply to

$$x = [\sigma(k+1,k+1) S_k]^{-1/2} s_k.$$

See Albert [1] for details.]

(9.1.8) **Exercise** (*Recursions for* S_k^{+})*:* Suppose $S_k \geqslant 0$ and

$$S_{k+1} = \begin{array}{c} \\ k \\ 1 \end{array}\!\!\begin{array}{c} \begin{array}{cc} k & \quad 1 \end{array} \\ \left(\begin{array}{c|c} S_k & s_k \\ \hline s_k^{\mathrm{T}} & \sigma(k+1,k+1) \end{array}\right) \end{array}$$

Let

$$t_k = S_k^{+} s_k \qquad \alpha_k = \sigma(k+1,k+1) - s_k^{\mathrm{T}} S_k^{+} s_k$$

$$\beta_k = 1 + \|t_k\|^2 \qquad T_k = I - t_k t_k^{\mathrm{T}}/\beta_k.$$

Then

(a) $S_{k+1} \geqslant 0$ if and only if $S_k t_k = s_k$ and $\alpha_k \geqslant 0$.
(b) In this case

$$S_{k+1}^{+} = \begin{cases} \begin{bmatrix} S_k^{+} + t_k t_k^{\mathrm{T}}/\alpha_k & -t_k/\alpha_k \\ -t_k^{\mathrm{T}}/\alpha_k & 1/\alpha_k \end{bmatrix} & \text{if} \quad \alpha_k > 0 \\ \begin{bmatrix} T_k S_k^{+} T_k & T_k S_k^{+} t_k/\beta_k \\ (T_k S_k^{+} t_k/\beta_k)^{\mathrm{T}} & t_k^{\mathrm{T}} S_k^{+} t_k/\beta_k^{2} \end{bmatrix} & \text{if} \quad \alpha_k = 0 \end{cases}$$

(see Albert [1]).

(9.2) Conditional Expectations for Normal Random Variables

A vector valued, normally distributed random variable $\mathbf{x}$ has a distribution which is completely specified by its mean,

$$m_{\mathbf{x}} \equiv \mathscr{E}\mathbf{x}$$

and its covariance

$$S_{\mathbf{xx}} = \mathscr{E}(\mathbf{x}-m_{\mathbf{x}})(\mathbf{x}-m_{\mathbf{x}})^{\mathrm{T}} \qquad \text{Anderson [1].}$$

It is known, Anderson [1], that if $\mathbf{y} = G\mathbf{x}$ and $\mathbf{z} = H\mathbf{x}$, then the conditional distribution of $\mathbf{y}$, given that $\mathbf{z} = z$ is, for each possible value of z [i.e., each $z \in \mathscr{R}(H)$] normal, and therefore specified completely by its conditional mean and covariance. We shall use the results of the last section to derive the key properties of conditional means and covariances in a streamlined fashion.

(9.2.1) **Theorem:** Suppose

$$\mathbf{x} = \begin{matrix} p \\ q \end{matrix}\begin{pmatrix} \mathbf{x}_1 \\ \hdashline \mathbf{x}_2 \end{pmatrix}$$

has a multivariate normal distribution with mean zero and covariance

$$S = \begin{matrix} p \\ q \end{matrix}\left(\begin{array}{c:c} S_{11} & S_{12} \\ \hdashline S_{12}^{\mathrm{T}} & S_{22} \end{array}\right).$$

Then

(a) $\mathbf{x}$ has the same distribution as

$$\mathbf{x}^{*} = \begin{matrix} p \\ q \end{matrix}\begin{pmatrix} \mathbf{x}_1^{*} \\ \hdashline \mathbf{x}_2^{*} \end{pmatrix}$$

where

$$\mathbf{x}_1{}^* = S_{11}^{1/2}\mathbf{w}_1$$
$$\mathbf{x}_2{}^* = S_{12}^{\mathrm{T}} S_{11}^{+} \mathbf{x}_1{}^* + (S_{22} - S_{12}^{\mathrm{T}} S_{11}^{+} S_{12})^{1/2}\mathbf{w}_2$$

and

$$\begin{matrix} p \\ q \end{matrix}\begin{pmatrix} \mathbf{w}_1 \\ \hline \mathbf{w}_2 \end{pmatrix}$$

has a normal distribution with mean zero and covariance I.

(b) The conditional distribution of $\mathbf{x}_2$ given that $\mathbf{x}_1 = x$ is normal with conditional mean

$$m_{2|1}(x) \equiv \mathscr{E}(\mathbf{x}_2 | \mathbf{x}_1 = x) = S_{12}^{\mathrm{T}} S_{11}^{+} x,$$

provided $x \in \mathscr{R}(S_{11})$.

The conditional covariance of $\mathbf{x}_2$ given that $\mathbf{x}_1 = x$ is

$$\begin{aligned} S_{22|1} &\equiv \mathscr{E}[(\mathbf{x}_2 - m_{2|1}(x))(\mathbf{x}_2 - m_{2|1}(x))^{\mathrm{T}} \,|\, \mathbf{x}_1 = x] \\ &= S_{22} - S_{12}^{\mathrm{T}} S_{11}^{+} S_{12} \end{aligned}$$

provided $x \in \mathscr{R}(S_{11})$. (Notice that $S_{22|1}$ does not depend explicitly on x.)

(c) If

$$\mathbf{x} = \begin{matrix} p \\ q \end{matrix}\begin{pmatrix} \mathbf{x}_1 \\ \hline \mathbf{x}_2 \end{pmatrix}$$

has mean value

$$m = \begin{matrix} p \\ q \end{matrix}\begin{pmatrix} m_1 \\ \hline m_2 \end{pmatrix}$$

which is different from zero, then the conditional covariance of $\mathbf{x}_2$ given that $\mathbf{x}_1 = x$ is the same as in (b) and

$$m_{2|1}(x) \equiv \mathscr{E}(\mathbf{x}_2 \,|\, \mathbf{x}_1 = x) = m_2 + S_{12}^{\mathrm{T}} S_{11}^{+}(x - m_1),$$

provided

$$x - m_1 \in \mathscr{R}(S_{11}).$$

Proof: (a) Since S is a covariance, it is nonnegative definite; so from (9.1.6), $S_{11} \geqslant 0$, $S_{22} - S_{12}^{\mathrm{T}} S_{11}^{+} S_{12} \geqslant 0$, hence both of the aforementioned matrices have square roots. Part (a) follows since $\mathbf{x}^*$ has a multivariate normal distribution with mean zero and (after a little algebra), covariance S. Notice that $\mathbf{x}_1{}^*$ (hence $\mathbf{x}_1$) must lie in $\mathscr{R}(S_{11}^{1/2}) = \mathscr{R}(S_{11})$ with probability 1.

(b) Without loss of generality, part (a) permits us to assume that $\mathbf{x}_1 = \mathbf{x}_1{}^*$ and $\mathbf{x}_2 = \mathbf{x}_2{}^*$ when we are exploring any properties of $\mathbf{x}$'s distribution.

Since $\mathbf{w}_1$ and $\mathbf{w}_2$ are independent, so then are $\mathbf{x}_1$ and $\mathbf{w}_2$ so that $\mathscr{E}(\mathbf{w}_2 \mid \mathbf{x}_1 = x) = \mathscr{E}(\mathbf{w}_2) = 0$. Thus, from part (a), if

$$x \in \mathscr{R}(S_{11}^{1/2}) = \mathscr{R}(S_{11})$$

$$m_{2|1}(x) = \mathscr{E}(\mathbf{x}_2 \mid \mathbf{x}_1 = x) = S_{12}^{\mathrm{T}} S_{11}^{+} \mathscr{E}(\mathbf{x}_1 \mid \mathbf{x}_1 = x) = S_{12}^{\mathrm{T}} S_{11}^{+} x$$

and

$$\begin{aligned} S_{22|1} &= \mathscr{E}[(\mathbf{x}_2 - m_{2|1}(x))(\mathbf{x}_2 - m_{2|1}(x))^{\mathrm{T}} \mid \mathbf{x}_1 = x] \\ &= \mathscr{E}[(T\mathbf{w}_2)(T\mathbf{w}_2)^{\mathrm{T}} \mid \mathbf{x}_1 = x] \end{aligned}$$

where

$$T = (S_{22} - S_{12}^{\mathrm{T}} S_{11}^{+} S_{12})^{1/2}.$$

Again, using the independence of $\mathbf{w}_2$ and $\mathbf{x}_1$, we see that

$$\begin{aligned} \mathscr{E}[(T\mathbf{w}_2)(T\mathbf{w}_2)^{\mathrm{T}} \mid \mathbf{x}_1 = x] &= \mathscr{E}(T\mathbf{w}_2 \mathbf{w}_2^{\mathrm{T}} T^{\mathrm{T}}) \\ &= TT^{\mathrm{T}} = (S_{22} - S_{12}^{\mathrm{T}} S_{11}^{+} S_{12}) \end{aligned}$$

since

$$\mathscr{E}(\mathbf{w}_2 \mathbf{w}_2^{\mathrm{T}}) = I.$$

(c) Apply (b) to $\mathbf{x}_2 - m_2$ and $\mathbf{x}_1 - m_1$:

$$\mathscr{E}(\mathbf{x}_2 - m_2 \mid \mathbf{x}_1 - m_1 = x - m_1) = S_{12}^{\mathrm{T}} S_{11}^{+}(x - m_1)$$

so

$$\mathscr{E}(\mathbf{x}_2 \mid \mathbf{x} = x) - m_2 = S_{12}^{\mathrm{T}} S_{11}^{+}(x - m_1).$$

The formula for the conditional covariance is unchanged if a constant vector is added to $\mathbf{x}_2 - m_2$. Again, we must tack on the restriction that $x - m_1$ be in $\mathscr{R}(S_{11})$ in order that the formulas hold. (See Marsaglia [1].) ■

The results of (9.2.1c) establish a relationship between the conditional mean and covariance of $\mathbf{u}$ given $\mathbf{v} = v$ and the unconditional mean and covariance of $\mathbf{u}$ for jointly normally distributed $\mathbf{u}$ and $\mathbf{v}$. The same argument can be applied to establish a similar relationship between the conditional mean (and covariance) of $\mathbf{u}$ given $\mathbf{v} = v$ and $\mathbf{y} = y$, and the conditional mean (and covariance) of $\mathbf{u}$ given $\mathbf{v} = v$, for any jointly normally distributed vector random variables, $\mathbf{v}$, $\mathbf{y}$, and $\mathbf{u}$:

(9.2.2) **Theorem:** Suppose

$$\begin{matrix} p+q \\ r \end{matrix} \begin{pmatrix} \mathbf{z} \\ \hline \mathbf{u} \end{pmatrix}$$

has a multivariate normal distribution with mean m and covariance

$$S = \left(\begin{array}{c|c} S_{\mathbf{zz}} & S_{\mathbf{zu}} \\ \hline S_{\mathbf{zu}}^{\mathrm{T}} & S_{\mathbf{uu}} \end{array}\right)$$

and suppose

$$\mathbf{z} = \begin{matrix} p \\ q \end{matrix}\left(\begin{array}{c} \mathbf{v} \\ \hline \mathbf{y} \end{array}\right).$$

Let

(9.2.2.1) $$m_{\mathbf{u}} = \mathscr{E}\mathbf{u} \qquad m_{\mathbf{v}} = \mathscr{E}\mathbf{v} \qquad m_{\mathbf{y}} = \mathscr{E}\mathbf{y}$$

(9.2.2.2) $$\begin{aligned} m_{\mathbf{z}|\mathbf{u}}(u) &= \mathscr{E}(\mathbf{z}\,|\,\mathbf{u}=u) \\ m_{\mathbf{v}|\mathbf{u}}(u) &= \mathscr{E}(\mathbf{v}\,|\,\mathbf{u}=u) \\ m_{\mathbf{y}|\mathbf{u}}(u) &= \mathscr{E}(\mathbf{y}\,|\,\mathbf{u}=u) \end{aligned}$$

and

(9.2.2.3) $$S_{\mathbf{zz}|\mathbf{u}} \qquad S_{\mathbf{vv}|\mathbf{u}} \qquad \text{and} \qquad S_{\mathbf{yy}|\mathbf{u}}$$

be the respective conditional covariances of $\mathbf{z}$, $\mathbf{v}$, and $\mathbf{y}$ given $\mathbf{u} = u$.

Then

(a) If $u - m_{\mathbf{u}} \in \mathscr{R}(S_{\mathbf{uu}})$, the conditional distribution of $\mathbf{z}$, given that $\mathbf{u} = u$, is normal with mean

(9.2.2.4) $$m_{\mathbf{z}|\mathbf{u}}(u) \equiv \begin{matrix} p \\ q \end{matrix}\left[\begin{array}{c} m_{\mathbf{v}|\mathbf{u}}(u) \\ \hline m_{\mathbf{y}|\mathbf{u}}(u) \end{array}\right] = m_{\mathbf{z}} + S_{\mathbf{zu}} S_{\mathbf{uu}}^{+}(u - m_{\mathbf{u}})$$

and covariance

(9.2.2.5) $$S_{\mathbf{zz}|\mathbf{u}} = \begin{matrix} \\ p \\ q \end{matrix}\begin{matrix} \quad p \qquad\quad q \\ \left(\begin{array}{c|c} S_{\mathbf{vv}|\mathbf{u}} & S_{\mathbf{vy}|\mathbf{u}} \\ \hline S_{\mathbf{vy}|\mathbf{u}}^{\mathrm{T}} & S_{\mathbf{yy}|\mathbf{u}} \end{array}\right) \end{matrix} = S_{\mathbf{zz}} - S_{\mathbf{zu}} S_{\mathbf{uu}}^{+} S_{\mathbf{zu}}^{\mathrm{T}}.$$

(b) If $u - m_{\mathbf{u}} \in \mathscr{R}(S_{\mathbf{uu}})$ and

$$y - m_{\mathbf{y}|\mathbf{u}}(u) \in \mathscr{R}(S_{\mathbf{yy}|\mathbf{u}})$$

then the conditional distribution of $\mathbf{v}$ given that $\mathbf{u} = u$ and $\mathbf{y} = y$, is normal with mean

(9.2.2.6) $$m_{\mathbf{v}|\mathbf{u},\mathbf{y}}(u, y) = m_{\mathbf{v}|\mathbf{u}}(u) + S_{\mathbf{vy}|\mathbf{u}} S_{\mathbf{yy}|\mathbf{u}}^{+}(y - m_{\mathbf{y}|\mathbf{u}}(u))$$

and covariance

(9.2.2.7) $$S_{\mathbf{vv}|\mathbf{u},\mathbf{y}} = S_{\mathbf{vv}|\mathbf{u}} - S_{\mathbf{vy}|\mathbf{u}} S_{\mathbf{yy}|\mathbf{u}}^{+} S_{\mathbf{vy}|\mathbf{u}}^{\mathrm{T}}$$

which does not depend explicitly on u and y.

Comment: Notice the formal similarity between (9.2.2.6) and (9.2.2.4) on the one hand and between (9.2.2.7) and (9.2.2.5), on the other.

Proof: Part (a) follows from (9.2.1) and from the fact that $\mathbf{z}$ can be partitioned in the form

$$\mathbf{z} = \begin{matrix} p \\ q \end{matrix}\begin{pmatrix} \mathbf{v} \\ \hline \mathbf{y} \end{pmatrix}.$$

(b) Let u be fixed and a *possible* value for the random variable $\mathbf{u}$ [i.e., $u - m_{\mathbf{u}} \in \mathscr{R}(S_{\mathbf{uu}})$]. The conditional joint distribution of $\mathbf{v}$ and $\mathbf{y}$ given that $\mathbf{u} = u$ is normal with mean

(9.2.2.8) $$\begin{pmatrix} m_{\mathbf{v}|\mathbf{u}}(u) \\ \hline m_{\mathbf{y}|\mathbf{u}}(u) \end{pmatrix} \equiv m_{\mathbf{z}|\mathbf{u}}(u)$$

and covariance

(9.2.2.9) $$\left(\begin{array}{c|c} S_{\mathbf{vv}|\mathbf{u}} & S_{\mathbf{vy}|\mathbf{u}} \\ \hline S_{\mathbf{vy}|\mathbf{u}}^{\mathrm{T}} & S_{\mathbf{yy}|\mathbf{u}} \end{array}\right) \equiv S_{\mathbf{zz}|\mathbf{u}}$$

where the elements of $m_{\mathbf{z}|\mathbf{u}}(u)$ and $S_{\mathbf{zz}|\mathbf{u}}$ are given in part (a). For fixed u, let

(9.2.2.10) $$\mathbf{z}^* = \begin{pmatrix} \mathbf{v}^* \\ \hline \mathbf{y}^* \end{pmatrix}$$

be a random variable which has a normal distribution with (unconditional) mean

(9.2.2.11) $$m_{\mathbf{z}^*} = m_{\mathbf{z}|\mathbf{u}}(u)$$

and (unconditional) covariance

(9.2.2.12) $$S_{\mathbf{z}^*\mathbf{z}^*} = S_{\mathbf{zz}|\mathbf{u}}.$$

Then for fixed u, the joint (unconditional) distribution of $\mathbf{v}^*$ and $\mathbf{y}^*$ is the same as the joint (conditional) distribution of $\mathbf{v}$ and $\mathbf{y}$, given that $\mathbf{u} = u$. Therefore, *the conditional distribution of* $\mathbf{v}^*$, *given that* $\mathbf{y}^* = y$ *is the same as the conditional distribution of* $\mathbf{v}$ *given that* $\mathbf{y} = y$ *and* $\mathbf{u} = u$ provided that y is an allowable realization of $\mathbf{y}^*$ [i.e., $y - m_{\mathbf{y}^*} \in \mathscr{R}(S_{\mathbf{y}^*\mathbf{y}^*})$].

So, for each y with $y - m_{\mathbf{y}^*} \in \mathscr{R}(S_{\mathbf{y}^*\mathbf{y}^*})$, the means and covariances of the two conditional distributions must coincide:

(9.2.2.13) $$m_{\mathbf{v}|\mathbf{u},\mathbf{y}}(u, y) = m_{\mathbf{v}^*|\mathbf{y}^*}(y)$$

(9.2.2.14) $$S_{\mathbf{vv}|\mathbf{u},\mathbf{y}} = S_{\mathbf{v}^*\mathbf{v}^*|\mathbf{y}^*}.$$

From part (a)

(9.2.2.15) $$m_{\mathbf{v}^*|\mathbf{y}^*}(y) = m_{\mathbf{v}^*} + S_{\mathbf{y}^*\mathbf{v}^*}^{\mathrm{T}} S_{\mathbf{y}^*\mathbf{y}^*}^{+}(y - m_{\mathbf{y}^*})$$

and

(9.2.2.16) $$S_{\mathbf{v}^*\mathbf{v}^*|\mathbf{y}^*} = S_{\mathbf{v}^*\mathbf{v}^*} - S_{\mathbf{v}^*\mathbf{y}^*} S^{+}_{\mathbf{y}^*\mathbf{y}^*} S^{\mathrm{T}}_{\mathbf{v}^*\mathbf{y}^*}.$$

Since

(9.2.2.17) $$m_{\mathbf{z}^*} = \begin{pmatrix} m_{\mathbf{v}^*} \\ \hdashline m_{\mathbf{y}^*} \end{pmatrix} = \begin{pmatrix} m_{\mathbf{v}|\mathbf{u}}(u) \\ \hdashline m_{\mathbf{y}|\mathbf{u}}(u) \end{pmatrix}$$ (9.2.2.11)

and

(9.2.2.18) $$S_{\mathbf{z}^*\mathbf{z}^*} = \left(\begin{array}{c:c} S_{\mathbf{v}^*\mathbf{v}^*} & S_{\mathbf{v}^*\mathbf{y}^*} \\ \hdashline S^{\mathrm{T}}_{\mathbf{v}^*\mathbf{y}^*} & S_{\mathbf{y}^*\mathbf{y}^*} \end{array}\right) = \left(\begin{array}{c:c} S_{\mathbf{vv}|\mathbf{u}} & S_{\mathbf{vy}|\mathbf{u}} \\ \hdashline S^{\mathrm{T}}_{\mathbf{vy}|\mathbf{u}} & S_{\mathbf{yy}|\mathbf{u}} \end{array}\right)$$ (9.2.2.12)

(9.2.2.6) follows from (9.2.2.13), (9.2.2.15), and (9.2.2.17) while (9.2.2.7) follows from (9.2.2.14), (9.2.2.16), and (9.2.2.18). ■

(9.2.2.19) **Exercise:** If

$$\begin{pmatrix} \mathbf{x} \\ \hdashline \mathbf{y} \\ \hdashline \mathbf{z} \end{pmatrix}$$

have a joint normal distribution and $\mathbf{z}$ is independent of

$$\begin{pmatrix} \mathbf{x} \\ \hdashline \mathbf{y} \end{pmatrix}$$

then $S_{\mathbf{xz}|\mathbf{y}} = 0$.

(9.2.3) **Exercise:** (a) For any matrices of the right size,

$$\operatorname{tr}(AB) = \operatorname{tr}(BA).$$

(b) If $\mathbf{x}$ is a random variable with covariance $S_{\mathbf{xx}}$, then $\mathscr{E}\|\mathbf{x}\|^2 = \operatorname{tr}(S_{\mathbf{xx}})$.

(9.2.4.1) **Exercise** (*Optimum properties of conditional expectation and wide-sense conditional expectation*): If

$$\begin{matrix} p \\ q \end{matrix}\begin{pmatrix} \mathbf{x} \\ \hdashline \mathbf{y} \end{pmatrix}$$

is a vector random variable with covariance matrix

$$\left(\begin{array}{c:c} S_{\mathbf{xx}} & S_{\mathbf{xy}} \\ \hdashline S^{\mathrm{T}}_{\mathbf{xy}} & S_{\mathbf{yy}} \end{array}\right)$$

we say that $m_{\mathbf{x}} + S_{\mathbf{xy}} S^{+}_{\mathbf{yy}}(y - m_{\mathbf{y}})$ is the *wide-sense conditional expectation* of $\mathbf{x}$ given $\mathbf{y} = y$. (If $\mathbf{x}$ and $\mathbf{y}$ have a joint normal distribution, the wide-sense conditional expectation coincides with the usual conditional expectation.)

(a) Assume that $m_{\mathbf{x}}$ and $m_{\mathbf{y}}$ are both zero. Show that $\mathscr{E}\|\mathbf{x}-A\mathbf{y}\|^2 \geqslant \mathscr{E}\|\mathbf{x}-S_{\mathbf{xy}}S_{\mathbf{yy}}^{+}\mathbf{y}\|^2$ for any matrix A of the right size, with strict inequality holding unless $AS_{\mathbf{yy}} = S_{\mathbf{xy}}$.

(b) If

$$\begin{matrix} p \\ q \end{matrix}\begin{pmatrix} \mathbf{x} \\ \hline \mathbf{y} \end{pmatrix}$$

has a joint normal distribution and $g(\cdot)$ is a measurable mapping from q-dimensional Euclidean space to p-dimensional Euclidean space, then

$$\mathscr{E}\|\mathbf{x}-g(\mathbf{y})\|^2 \geqslant \mathscr{E}\|\mathbf{x}-S_{\mathbf{xy}}S_{\mathbf{yy}}^{+}\mathbf{y}\|^2$$

with strict inequality holding unless $g(y) = S_{\mathbf{xy}}S_{\mathbf{yy}}^{+}y$ for almost all y.

Comment: (a) Shows that the wide sense conditional expectation is the best linear predictor for $\mathbf{x}$, based on $\mathbf{y}$ when mean square error is to be minimized.

(b) Shows that the wide sense conditional expectation is the best predictor of all (linear or not), in the normal case.

(9.2.4.2) **Exercise:** Suppose

$$S_{\mathbf{zz}} = \begin{matrix} \\ p \\ q \end{matrix}\begin{matrix} \begin{matrix} p & q \end{matrix} \\ \begin{pmatrix} S_{\mathbf{uu}} & S_{\mathbf{uy}} \\ S_{\mathbf{uy}}^{\mathrm{T}} & S_{\mathbf{yy}} \end{pmatrix} \end{matrix} \geqslant 0.$$

Then $u \in S_{\mathbf{uu}}$ and $y - S_{\mathbf{uy}}^{\mathrm{T}}S_{\mathbf{uu}}^{+}u \in \mathscr{R}(S_{\mathbf{yy}|\mathbf{u}})$ if and only if

$$z = \begin{pmatrix} u \\ \hline y \end{pmatrix} \in \mathscr{R}(S_{\mathbf{zz}}).$$

Comment: The last result explains away the apparently asymmetric conditions on $u - m_{\mathbf{u}}$ and $y - m_{\mathbf{y}|\mathbf{u}}(u)$ which are necessary in order that (9.2.2b) should hold. In short, (9.2.2.6) and (9.2.2.7) hold provided

$$\begin{pmatrix} u - m_{\mathbf{u}} \\ \hline y - m_{\mathbf{y}} \end{pmatrix} \in \mathscr{R}(S_{\mathbf{zz}}).$$

(9.2.5) **Exercise:** Suppose

$$\begin{pmatrix} \mathbf{x}^* \\ \hline \mathbf{w} \\ \hline \mathbf{y} \end{pmatrix}$$

has a multivariate normal distribution. Let

$$\begin{aligned} m_{\mathbf{x}^*|\mathbf{y}}(y) &= \mathscr{E}(\mathbf{x}^*|y=y) \\ m_{\mathbf{w}} &= \mathscr{E}(\mathbf{w}) \\ S_{\mathbf{x}^*\mathbf{x}^*|\mathbf{y}} &= \mathscr{E}[(\mathbf{x}^*-m_{\mathbf{x}^*|\mathbf{y}}(y))(\mathbf{x}^*-m_{\mathbf{x}^*|\mathbf{y}}(y))^{\mathrm{T}}\,|\,\mathbf{y}=y] \end{aligned}$$

and

$$S_{\mathbf{ww}} = \mathscr{E}(\mathbf{w}-m_{\mathbf{w}})(\mathbf{w}-m_{\mathbf{w}})^{\mathrm{T}}.$$

If

(9.2.5.1) $\mathbf{x} = \Phi\mathbf{x}^* + \mathbf{w}$ (where Φ is a deterministic *square matrix*)

and if $\mathbf{w}$ is independent of

$$\begin{pmatrix} \mathbf{x}^* \\ --- \\ \mathbf{y} \end{pmatrix}$$

then

(9.2.5.2) $$m_{\mathbf{x}|\mathbf{y}}(y) = \Phi m_{\mathbf{x}^*|\mathbf{y}}(y) + m_{\mathbf{w}}$$

(9.2.5.3) $$S_{\mathbf{xx}|\mathbf{y}} = \Phi S_{\mathbf{x}^*\mathbf{x}^*|\mathbf{y}}\Phi^{\mathrm{T}} + S_{\mathbf{ww}}.$$

(9.2.6) **Exercise:** If $\mathbf{v}$ has a normal distribution with mean 0 and covariances $S_{\mathbf{vv}}$ and is independent of

$$\begin{pmatrix} \mathbf{x} \\ -- \\ \mathbf{y} \end{pmatrix}$$

which has a normal distribution, and if

(9.2.6.1) $$\mathbf{z} = H\mathbf{x} + \mathbf{v}$$

where H is a deterministic matrix, then

(9.2.6.2) $$m_{\mathbf{x}|\mathbf{yz}}(y,z) = m_{\mathbf{x}|\mathbf{y}}(y) + S_{\mathbf{xx}|\mathbf{y}}H^{\mathrm{T}}(HS_{\mathbf{xx}|\mathbf{y}}H^{\mathrm{T}}+S_{\mathbf{vv}})^{+}(z-Hm_{\mathbf{x}|\mathbf{y}}(y))$$

and

(9.2.6.3) $$S_{\mathbf{xx}|\mathbf{yz}} = S_{\mathbf{xx}|\mathbf{y}} - S_{\mathbf{xx}|\mathbf{y}}H^{\mathrm{T}}(HS_{\mathbf{xx}|\mathbf{y}}H^{\mathrm{T}}+S_{\mathbf{vv}})^{+}HS_{\mathbf{xx}|\mathbf{y}}$$

where $m_{\mathbf{x}|\mathbf{y},\mathbf{z}}(y,z)$ is the conditional expectation of $\mathbf{x}$ given that $\mathbf{y}=y$ and $\mathbf{z}=z$, $S_{\mathbf{xx}|\mathbf{y},\mathbf{z}}$ is the conditional covariance of $\mathbf{x}$ given $\mathbf{y}=y$ and $\mathbf{z}=z$, etc.

Comment: (9.2.6) shows how the acquisition of new data changes the *posterior* distribution. If $\mathbf{y}$ represents all the data up to the last sampling period, then the *posterior* distribution of $\mathbf{x}$ (i.e., conditional distribution) given the data $\mathbf{y}=y$, contains all of the available information about $\mathbf{x}$ that can be learned from analyzing $\mathbf{z}$. This distribution, in the normal case, is characterized by its mean and covariance. If a new datum of the form (9.2.6.1) is observed, then the posterior distribution of $\mathbf{x}$ changes and is now characterized by its conditional mean and covariance given $\mathbf{y}$ *and* $\mathbf{z}$. (9.2.6.2) and (9.2.6.3) show how the new mean and covariance are related to the old.

(9.3) Kalman Filtering

Suppose $\{\mathbf{x}_n;\ n = 0, 1, 2, \ldots\}$ is a vector valued stochastic process whose evolution is governed by a so-called "state equation" of the form

(9.3.1) $$\mathbf{x}_{n+1} = \Phi_n \mathbf{x}_n + \mathbf{w}_n$$

where $\{\mathbf{w}_n;\ n = 0, 1, 2, \ldots\}$ is a sequence of independent vector valued normal random variables with means 0 and covariances

(9.3.2) $$T_n = \mathscr{E}\mathbf{w}_n \mathbf{w}_n^{\mathrm{T}}.$$

The so-called "state variable" can be partially observed in the presence of noise, the observations being of the form

(9.3.3) $$\mathbf{z}_n = H_n \mathbf{x}_n + \mathbf{v}_n \qquad n = 1, 2, \ldots$$

where $\{\mathbf{v}_n;\ n = 1, 2, \ldots\}$ is a sequence of independent vector valued normal random variables with zero means and covariances

(9.3.4) $$R_n = \mathscr{E}\mathbf{v}_n \mathbf{v}_n^{\mathrm{T}}.$$

The $\{\mathbf{v}_n\}$ and $\{\mathbf{w}_n\}$ processes are assumed to be independent,[1] and the initial state $\mathbf{x}_0$ is assumed to be a normally distributed zero mean random vector which is independent of the $\mathbf{v}_n$'s and $\mathbf{w}_n$'s. Under these assumptions, it is not hard to show that the joint distribution of $\mathbf{x}_0, \mathbf{x}_1, \ldots, \mathbf{x}_{n+1}, \mathbf{z}_1, \ldots, \mathbf{z}_n$ is normal.

In (9.2.4.1), it was established that the best mean square predictor for $\mathbf{x}_n$, based upon the observations

$$\mathbf{z}_1 = z_1,\ \mathbf{z}_2 = z_2,\ \ldots, \mathbf{z}_n = z_n$$

is

(9.3.5) $$x_{n|n} \equiv \mathscr{E}(\mathbf{x}_n \mid \mathbf{z}_1 = z_1, \ldots, \mathbf{z}_n = z_n)$$

where we adopt the convention that

$$x_{0|0} = 0.$$

[1] This assumption can be dispensed with at the expense of slightly more complicated formulas.

Similarly the best predictor for $\mathbf{x}_{n+1}$ given the first n observations, is

(9.3.6) $$x_{n+1|n} = \mathscr{E}(\mathbf{x}_{n+1} \mid \mathbf{z}_1 = z_1, \ldots, \mathbf{z}_n = z_n).$$

If we denote the conditional covariances of $\mathbf{x}_n$ and $\mathbf{x}_{n+1}$, given $\mathbf{z}_1 = \mathbf{z}_1, \ldots, \mathbf{z}_n = z_n$ by $S_{n|n}$ and $S_{n+1|n}$ respectively, then by virtue of (9.2.5) and (9.2.6)

(9.3.7) $$x_{n+1|n} = \Phi_n x_{n|n} \qquad n = 0, 1, 2, \ldots$$

(9.3.8) $$x_{n+1|n+1} = x_{n+1|n} + K_{n+1}(z_{n+1} - H_{n+1} x_{n+1|n})$$

where

(9.3.9) $$K_{n+1} = S_{n+1|n} H_{n+1}^{\mathrm{T}} (H_{n+1} S_{n+1|n} H_{n+1}^{\mathrm{T}} + R_{n+1})^{+}$$

(9.3.10) $$S_{n+1|n} = \Phi_n S_{n|n} \Phi_n^{\mathrm{T}} + T_n$$

and

(9.3.11) $$S_{n+1|n+1} = S_{n+1|n} - S_{n+1|n} H_{n+1}^{\mathrm{T}} \times (H_{n+1} S_{n+1|n} H_{n+1}^{\mathrm{T}} + R_{n+1})^{+} H_{n+1} S_{n+1|n}.$$

These relations are initialized by

(9.3.12) $$x_{0|0} = 0 \qquad S_{0|0} = \mathscr{E}\mathbf{x}_0 \mathbf{x}_0^{\mathrm{T}}.$$

[These recursions follow instantly from (9.2.5) and (9.2.6) if we identify

$$\mathbf{y} \text{ with the ``super vector''} \begin{bmatrix} \mathbf{z}_1 \\ \hline \mathbf{z}_2 \\ \hline \vdots \\ \hline \mathbf{z}_n \end{bmatrix}$$

$\mathbf{z}$ with $\mathbf{z}_{n+1}$, $\mathbf{x}^*$ with $\mathbf{x}_n$, and $\mathbf{x}$ with $\mathbf{x}_{n+1}$.]

The relations (9.3.7)–(9.3.12) are the famous Kalman "filtering equations," Kalman [1], about which an enormous literature has grown up in the systems engineering journals. The practical importance of these recursions is best exemplified by Battin [1], Battin and Levine [1], and American Statistical Association [1].

Comments: The restrictions that $\mathbf{x}_0$ and the state equation noise have zero means is not essential. If $\mathbf{x}_0$ has mean m_0, then the recursion (9.3.7) is reinitialized by $x_{0|0} = m_0$. If the state equation noise, $\mathbf{w}_n$, has mean m_n, then (9.3.7) becomes

(9.3.7′) $$x_{n+1|n} = \Phi_n x_{n|n} + m_n.$$

The key feature of the Kalman equations are their recursive nature. The stream of data $\mathbf{z}_1, \mathbf{z}_2, \ldots$ is incorporated "in real time" to produce a stream of estimates for the current values of the state variable.

The normality assumption can be dropped, in which case, the recursions generate wide-sense conditional expectations.

(9.4) The Relationship between Least Squares Estimates and Conditional Expectations

Suppose

(9.4.1)
$$\mathbf{z} = H\mathbf{x} + \mathbf{v}$$

where

$$\begin{matrix} p \\ n \end{matrix}\begin{pmatrix} \mathbf{x} \\ \hline \mathbf{v} \end{pmatrix}$$

has a normal distribution with mean 0 and covariance

$$S = \begin{matrix} \\ p \\ n \end{matrix}\begin{matrix} \begin{matrix} p & & n \end{matrix} \\ \left(\begin{array}{c|c} (\lambda^{-2})^{-1}C^2 & 0 \\ \hline 0 & \mathrm{I} \end{array}\right) \end{matrix}.$$

The conditional expectation of $\mathbf{x}$ given $\mathbf{z} = z$ is

(9.4.2)
$$m_{\mathbf{x}|\mathbf{z}}(z) = \lambda^{-2}C^2H^{\mathrm{T}}(\lambda^{-2}HC^2H^{\mathrm{T}} + I)^{-1}z.$$

By (4.9.5),

(9.4.3)
$$\left(\frac{HC^2H^{\mathrm{T}}}{\lambda^2} + I\right)^{+} = I - (HC^2H^{\mathrm{T}})(HC^2H^{\mathrm{T}})^{+} + \lambda^2(HC^2H^{\mathrm{T}})^{+} + O(\lambda^4) \quad \text{as} \quad \lambda \to 0$$
$$= I - (HC)(HC)^{+} + \lambda^2(HC^2H^{\mathrm{T}})^{+} + O(\lambda^4) \; (3.11.7)$$

so that

$$m_{\mathbf{x}|\mathbf{z}}(z) = (C/\lambda^2)[(I - (HC)(HC)^{+})(HC)]^{\mathrm{T}}z + C(HC)^{+}(HC^2H^{\mathrm{T}})^{+}z + O(\lambda^2)$$
$$= C(HC)^{+}z + O(\lambda^2) \qquad \text{as} \quad \lambda \to 0.$$

But $C(HC)^{+}z$ minimizes $\|z - Hx\|^2$ subject to the constraint $x \in \mathscr{R}(C)$ since

$$\min_{x \in \mathscr{R}(C)} \|z - Hx\|^2 = \min_{y} \|z - HCy\|^2$$

occurs when $x = C\hat{y}$ and $\hat{y} = (HC)^{+}z$. Thus, for small values of λ, $\mathscr{E}(\mathbf{x} \mid \mathbf{z} = z)$ is close to the constrained least squares estimator for $\mathbf{x}$ [subject to the constraint $(I - CC^{+})x = 0$].

If C is nonsingular, $I-CC^{+}=0$ and so in this case, $\mathscr{E}(\mathbf{x}\,|\,\mathbf{z}=z)$ is close to the naive least squares estimate for $\mathbf{x}$.

(9.4.4) **Example:** Suppose $\{\mathbf{v}_n;\ n=1,2,\ldots\}$ is a sequence of independent normally distributed random variables with mean zero and variances $\sigma_n{}^2$. Suppose, $\mathbf{x}$ is a zero mean p-dimensional vector random variable having a normal distribution with mean 0 and covariance $(1/\lambda^2)C$. Assume that $\mathbf{x}$ is independent of the $\mathbf{v}_n$'s.

Suppose observations of the form

$$\boldsymbol{\zeta}_n = h_n{}^{\mathrm{T}}\mathbf{x} + \mathbf{v}_n \qquad n = 1,2,\ldots$$

are made in order to estimate x. Let $\hat{x}_n(\lambda)$ be the conditional expectation of $\mathbf{x}$ given $\boldsymbol{\zeta}_1=\zeta_1,\ldots,\boldsymbol{\zeta}_n=\zeta_n$, and let $S_n(\lambda)$ be the conditional covariance of $\mathbf{x}_n$ given the data up through the nth.

Then the Kalman recursions reduce to

$$\hat{x}_0(\lambda) = 0 \qquad S_0(\lambda) = \lambda^{-2}C,$$

$$\hat{x}_{n+1}(\lambda) = \hat{x}_n(\lambda) + \frac{S_n(\lambda)\,h_{n+1}}{\sigma_{n+1}^2 + h_{n+1}^{\mathrm{T}}\,S_n(\lambda)\,h_{n+1}}(\zeta_{n+1} - h_{n+1}^{\mathrm{T}}\,x_n(\lambda))$$

$$S_{n+1}(\lambda) = S_n(\lambda) - \frac{(S_n(\lambda)\,h_{n+1})(S_n(\lambda)\,h_{n+1})^{\mathrm{T}}}{\sigma_{n+1}^2 + h_{n+1}^{\mathrm{T}}\,S_n(\lambda)\,h_{n+1}}.$$

Compare this with the recursions (8.3.1) and (8.4.5). [Read B_n for $S_n(\lambda)$.] When λ is close to zero, the present recursions generate a sequence which is close to $\bar{x}_n$, the least squares estimate for x based on $\zeta_1,\ldots,\zeta_n$ computed subject to the constraints $(I-CC^{+})x=0$. (See Albert and Sittler [1].)

REFERENCES

Albert, A.

[1] Conditions for positive and nonnegative definiteness in terms of pseudoinverses. *SIAM J. Appl. Math.* **17** (1969), 434–440.

Albert, A., and Sittler, R.

[1] A method for computing least squares estimators that keep up with the data. *SIAM J., Control* **3** (1965), 394–417.

American Statistical Association

[1] Regression procedures for missile trajectory estimation. *Proc. of the 105th Regional Meeting*, Florida State Univ. (1965).

Anderson, T. W.

[1] "An Introduction to Multivariate Statistical Analysis." Wiley, New York, 1958.

Battin, R. H.

[1] "Astronautical Guidance." McGraw-Hill, New York, 1964.

Battin, R. H., and Levine, G.

[1] Application of Kalman filtering techniques to the Apollo program. MIT Inst. Lab. Tech. Rep. E2401, April 1969.

Bellman, R.

[1] "Introduction to Matrix Analysis." McGraw-Hill, New York 1960.

Ben-Israel, A.

[1] On error bounds for generalized inverses. *SIAM J. Numer. Anal.* **3** (1966), 585–592.

Ben-Israel, A., and Charnes, A.

[1] Contribution to the theory of generalized inverses. *SIAM J.* **11** (1963), 667–699.

[2] An explicit solution of a special class of linear programming problems. *Operations Res.* **16** (1968), 1166–1175.

Ben-Israel, A., Charnes, A., and Robers, P. D.

[1] On generalized inverses and interval linear programming. In "Theory and Application of Generalized Inverses" (T. Boullion and P. Odell, eds.). Proceedings of a symposium at Texas Technological College, March, 1968.

Ben-Israel, A. and Ijiri, Y.

[1] A report on the machine calculation of the generalized inverse of an arbitrary matrix. *ONR Research Memo No. 110*, Carnegie Inst. Of Tech., March 1963.

Ben-Israel, A., and Robers, P. D.

[1] A suboptimization method for interval linear programming. *Systems Res. Memo No. 204*, Northwestern Univ., June 1968.

Ben-Israel, A., and Wersan, S. J.

[1] An elimination method for computing the generalized inverse for arbitrary complex matrix. *J. Assoc. Comput. Mach.* **10** (1963), 532–537.

Bocher, M.

[1] "Introduction to Higher Algebra." Macmillan, New York, 1907.

Boullion, T. and Odell, P. (eds.)

[1] "Theory and Application of Generalized Inverses." Proceedings of symposium at Texas Technological College, March 1968.

[2] "Generalized Inverse Matrices." Wiley (Interscience), New York, 1971.

den Broeder, G. G., and Charnes, A.

[1] Contributions to the theory of generalized inverses for matrices. *ONR Res. Memo No. 39*. Northwestern Univ., 1962.

Butler, T., and Martin, A. V.

[1] On a method of Courant for minimizing functionals. *J. Math. Phys.* **41** (1962), 291–299.

Cline, R. E.

[1] Note on the generalized inverse of the product of matrices. *SIAM Rev.* **6** (1964), 57–58.

[2] Representations for the generalized inverse of a partitioned matrix. *SIAM J. Appl. Math.* **12** (1964), 588–600.

[3] Representations for the generalized inverse of sums of matrices. *SIAM J. Numer. Anal.* **2** (1965), 99–114.

Decell, H. P.

[1] An alternate form of the generalized inverse of an arbitrary complex matrix. *SIAM Rev.* **7** (1965), 356–358.

[2] An application of the Cayley–Hamilton Theorem to generalized matrix inversion. *SIAM Rev.* **7** (1965), 526–528.

Decell, H. P., and Odell, P. L.

[1] On the fixed point probability vector of regular or ergodic transition matrices. *J. Am. Stat. Assoc.* **62** (1967), 600–602.

Fadeev, D. K., and Fadeeva, V. N.

[1] "Computational Methods of Linear Algebra." Freeman, San Francisco, 1963.

Feller, W.

[1] "An Introduction to Probability Theory and Its Applications," 3rd ed., Vol. I. Wiley, New York, 1952.

Goldman, A. J., and Zelen, J.

[1] Weak generalized inverses and minimum variance linear unbiased estimation. *J. Res. Nat. Bur. Standards Sect. B*, **68B** (1964), 151–172.

Golub, G.
[1] Numerical methods for solving linear least squares problems. *Numer. Math.* **7** (1965), 206–216.
[2] Least squares singular values and matrix approximations. *Apl. Mat.* (Prague) **13** (1968), 44–51.

Golub, G., and Kahan, W.
[1] Calculating the singular values and pseudo-inverse of a matrix. *SIAM J. Numer. Anal.* **2** (1965), 205–224.

Good, I. J.
[1] Some applications of the singular decomposition of a matrix. *Technometrics* **11** (1969), 823–831.

Greville, T. N. E.
[1] The pseudoinverse of a rectangular matrix and its applications to the solution of systems of linear equations. *SIAM Rev.* **1** (1959), 38–43.
[2] Some applications of the pseudoinverse of a matrix. *SIAM Rev.* **2** (1960), 15–22.
[3] Note on the generalized inverse of a matrix product. *SIAM Rev.* **8** (1966), 518–521. [Erratum **9** (1967).]

Halmos, Paul R.
[1] Finite dimensional vector spaces. *Ann. of Math. Studies* **7** (1955).

Kalman, R. E.
[1] A new approach to linear filtering and prediction problems. *J. Basic Engrg.* **82** (1960), 35–45.

Karlin, S.
[1] "Mathematical Methods and Theory in Games Programming and Economics." Addison-Wesley, Reading, Massachusetts, 1959.

Kruskal, W.
[1] When are the Gauss–Markov and least squares estimators identical? A coordinate free approach. *Ann. Math. Statist.* **39** (1968), 70–75.

Marsaglia, G.
[1] Conditional means and covariances of normal variables with singular covariance matrices. *J. Am. Stat. Assoc.* **59** (1965), 1203–1204.

Mitra, S. K., and Rao, C. R.
[1] Conditions for optimality and validity of simple least squares theory. *Ann. Math. Statist.* **40** (1968), 1617–1624.

Moore, E. H.
[1] Abstract. *Bull. Amer. Math. Soc.* **26** (1920), 394–395.
[2] General analysis, Part I. *Memoirs Amer. Philos. Soc.* **1** (1935), 1–231.

Noble, B.
[1] A method for computing the generalized inverse of a matrix. *SIAM J. Numer. Anal.* **3** (1966), 582–584.

Osborne, E. E.
[1] Smallest least squares solutions of linear equations. *SIAM J. Numer. Anal.* **2** (1965), 300–307.

Penrose, R.
[1] A generalized inverse for matrices. *Proc. Cambridge Philos. Soc.* **51** (1955), 406–413.
[2] On best approximate solutions of linear matrix equations. *Proc. Cambridge Philos. Soc.* **52** (1956), 17–19.

Pereyre, V., and Rosen, J. B.
[1] Computation of the pseudoinverse of a matrix of unknown rank. Computer Sciences Division, Stanford Univ., Tech. Rep. CS 13, Sept. (1964).
Price, C. M.
[1] The matrix pseudoinverse and minimal variance estimates. *SIAM Rev.* **6** (1964), 115–120.
Pyle, L. D.
[1] Generalized inverse computations using the gradient projection method. *J. Assoc. Comput. Mach.* **11** (1964), 422–429.
Rao, C. R.
[1] A note on a generalized inverse of a matrix with applications to problems in mathematical statistics. *J. Roy. Statist. Soc. Ser. B*, **24** (1962), 152–158.
Rao, C. R. and Mitra, S. K.
[1] "Generalized Inverse of Matrices and Its Application." Wiley, New York, 1971.
Rust, B., Burrus, W. R., and Schneeberger, C.
[1] A simple algorithm for computing the generalized inverse of a matrix. *Comm. ACM* **9** (1966), 381–386.
Scheffé, H.
[1] "The Analysis of Variance." Wiley, New York, 1959.
Stewart, G. W.
[1] On the continuity of the generalized inverse. *SIAM J. Appl. Math.* **17** (1969), 33–45.
Tewarson, R. P.
[1] A direct method for generalized matrix inversion. *SIAM J. Numer. Anal.* **4** (1967), 499–507.
[2] On two direct methods for computing generalized inverses. *Computing*, **7** (1971), 236–239.
Watson, G. S.
[1] Linear least squares regression. *Ann. Math. Statist.* **38** (1967), 1679–1699.
Zyskind, G.
[1] On canonical forms, nonnegative covariance matrices and best and simple least squares linear estimators in linear models. *Ann. Math. Statist.* **38** (1967), 1092–1109.
Zyskind, G., and Martin, F. B.
[1] On best linear estimation and a general Gauss–Markov theorem in linear models with arbitrary nonnegative covariance structure. *SIAM J. Appl. Math.* **17** (1969), 1190–1202.

INDEX

Numbers in italics refer to the pages on which the complete references are listed.

A

Adjoint, 10
Albert, A., 158, 161, 172, *173*
Analysis of variance, 108
 table, 109
Anderson, T. W., 94, 161, *173*

B

Battin, R. H., 170, *173*
Bellman, R., 5, 12, 75, 158, *173*
Ben-Israel, A., 33, 42, 65, 69, 74, *173*, *174*
BLUE (best linear unbiased estimator), 88 ff., 105, 126, 148 ff.
 and constraints, 122
 for estimable vector parametric functions, 97 ff.
 and naive least squares, 92
 and stepwise regression, 117
 and straight line regression, 113
 and two-way layout, 116
Bocher, M., 78, *174*
Boullion, T., *174*
Bounded linear program, 33
den Broeder, G. G., 20, *174*
Burrus, W. R., 57, *176*
Butler, T., 120, *174*

C

Cayley–Hamilton theorem, 75
Characteristic polynomial, 76
 computation of coefficients, 76–77
Charnes, A., 20, 33, 42, 74, *173*, *174*
Chi-square distribution, 94 ff., 98 ff. 109
 and projections, 97, 100
Cline, R. E., 49, 55, *174*
Cochran's theorem, 52
Computational methods,
 based on Cayley–Hamilton theorem,74 ff.
 based on Gauss–Jordan elimination, 65 ff.
 based on gradient projection, 69 ff.
 based on Gramm–Schmidt orthogonalization, 57 ff.
Conditional covariance,
 normal case, 161
 recursion for, 164
Conditional distribution, normal case, 161
Conditional expectation,
 and least squares, 171
 as optimal mean square predictor, 166
 for normal case, 161
 recursion for, 164
Confidence ellipsoids, 97 ff.
 and tests of general linear hypothesis, 102
Constrained least squares, 31, 121
 recursive computation of, 133

Constraint set for linear program, 33
Constraints, as fictitious observations, 135
Covariance of BLUE, recursive computation of, 127

D

Decell, H. P., 38, 74, *174*
Degrees of freedom, 94 ff., 98 ff., 109
Dependent variable (in regression), 86
Diagonalization theorem, 12, 27
Differential correction, 127
Dimension of linear manifold, 52
Dominant eigenvalue algorithm, 42

E

Eigenvalue, 12, 23, 24, 38 ff.
Eigenvector, 24, 38 ff.
Estimable, 88
Estimable parametric function, 88 ff.
Estimable vector parametric functions, 97 ff.
EVPF (estimable vector parametric function), 97 ff.

F

F distribution, 94 ff., 99 ff., 101–102
Fadeev, D. K., 74, 76, *174*
Fadeeva, V. N., 74, 76, *174*
Feasible linear program, 33
Feller, W., 37, *174*
Fourier expansion, 8

G

Gauss–Jordan elimination, 65
Gauss–Markov estimator, 97
Gauss–Markov theorem, generalized, 90
General linear hypothesis, tests of, 100 ff.
General linear model, 87
Goldman, A. J., 123, *174*
Golub, G., 42, 81, *175*
Good, I. J., 38, *175*
Gramm–Schmidt orthogonalization, 9, 57, 58 ff., 70 ff., 112
  modified, 81
  transformation expressed as a matrix, 60
Grammiam, 12
Greville, T. N. E., 43, 53, *175*

H

Halmos, P. R., 5, *175*

I

Ijiri, Y., 69, *174*
Independent variable (in regression), 86
Inner product, 5
Inverse, 12

K

Kahan, W., 42, 81, *175*
Kalman, filtering, 169 ff.
Kalman, R. E., 170, *175*
Karlin, S., 5, *175*
Kruskal, W., 93, *175*

L

Least squares estimator, recursive computation of, 125 ff.
Least squares,
  and conditional expectation, 171
  generalized, 93
  minimum norm solution to, 15, 17
Levine, G., 170, *173*
Likelihood ratio test, 100, 104
Linear equations, 30
Linear manifold, 6 ff.
  definition of, 5
Linear programming, 33
Linearly independent, 12, 21

M

Markov chain, 37
Marsaglia, G., 163, *175*
Martin, A. V., 120, *174*
Martin, F. B., 90, 123, *176*
Matrix equations, 35–36
Missing observations,
  general case, 153
  in orthogonal designs, 151
Mitra, S., 93, *175*, *176*
Modified Gramm–Schmidt orthogonalization, 81
Moore, E. H., 3, 20, *175*
Moore–Penrose generalized inverse, definition of, 20
Multiple correlation coefficient, 46

N

Naive least squares and BLUE, 92
Noble, B., 65, *175*
Noncentrality parameter, 94 ff., 97
Nonnegative definiteness, 157 ff.

Nonsingular, 12
Norm, 5
Normal equations, 17
Normal matrix, 36
Normal random variables, 94
Null space, 10 ff., 15

O

Odell, P. L., 38, *174*
Orthogonal, 5
  to a linear manifold, 6
Orthogonal complement of a linear manifold, definition of, 11
Orthogonal design, 104 ff.
  and constrained least squares, 105
  and projections, 106
  and sum of squares decomposition, 108
Orthogonal matrix, 12, 36
Orthonormal, 24
Osborne, E. E., *175*

P

Parametric function, 87, 88
Partial correlation coefficient, 46
Partial isometry, 41
Penalty functions, 119 ff.
Penrose, R., 20, 42, *175*
Penrose conditions, 28
Permutation matrix, 57
Perturbation theorems, 50 ff.
Pereyre, 81, *176*
Positive definiteness, 157 ff.
Price, C. M., 93, *176*
Probability vector, 37
Projection, 8, 16 ff., 18 ff.
  on null space of matrix, 20, 42, 47
  on range of matrix, 20
  recursion for, 37
Projection matrix,
  definition of, 26
  eigenvalues of, 26
Projection theorem, 6, 11
Projections and pseudoinverse, 20
Pseudoinverse,
  and constrained least squares, 31
  definition of, 19
  discontinuous nature of, 25
  and linear programming, 33
  and matrix equations, 35
  Penrose conditions, 28–29
  and perturbation theorems, 50 ff.
  of products, 29, 52 ff.
  and projections, 20
  properties, 26 ff., 30
  recursion for nonnegative definite matrix, 47, 49
  recursion for partitioned matrix, 43–44, 49
  recursive computation for nonnegative definite matrix, 160
  and roundoff, 25
  for special case, 22 ff.
  and theory of linear equations, 30
Pyle, L. D., 69, *176*

Q

Quadratic forms, distribution theory for normal case, 93

R

Range space, 10, 11, 12, 15, 17
Rank of a matrix, 52
Rao, C. R., 93, 96, *175*, *176*
Recursion, 48
  for conditional covariance, 164
  for conditional expectation, 164
  for constrained least squares estimator, 133
  for covariance of BLUE, 127
  for least squares estimator, observations missing, 153 ff.
  for projections, 37
  for pseudoinverse of nonnegative definite matrix, 47, 49, 160
  for pseudoinverse of partitioned matrix, 43–44, 49
  for residual error, 129
  in weighted least squares, 132
  for steady state probability of Markov chain, 37
  for straight line regression, 142
  for unconstrained least squares, 125 ff.
  for weighted least squares estimator, 131 ff.
Regression
  multiple, 46
  stepwise, 44 ff.
  straight line, 112

Residual covariance, 88
Residual error,
  recursive computation, for weighted least squares, 132
  recursive computation of, 129
Residual sum of squares, for analysis of covariance, 150
Robers, P. D., 33, *174*
Rosen, J. B., 81, *176*
Rust, B., 57, 63, *176*

S

Scalar product, 5
Scheffé, H., 52, 87, 94, 109, 148, *176*
Schneeburger, C., 57, *176*
Singular, 12
Singular value decomposition theorem, 38–39
Sittler, R., 172, *173*
Smoothing vector, 127
Spanning vectors, 7
  for linear manifold, 8
Spectral representation, 24
Steady state probability vector of Markov chain, recursion for, 37
Stepwise regression, 44 ff., 117, 146
  and analysis of covariance, 148
Stewart, G. W., 51, *176*
Stochastic matrix, 37
Straight line regression, 112
  recursive computation of, 142
Sum of squares decomposition, 108, 110
  for straight line regression, 114
  for two-way layout, 116
Symmetric matrix, 11 ff.

T

Tests of general linear hypothesis, and confidence ellipsoids, 102
Tewarson, R. P., 69, *176*
Transpose, 5
TRUE-BLUE, 93
Two-way layout, 115 ff.
Typographical conventions, 4

U

Unbiased, 88

V

Variance, 88

W

Watson, G. S., 93, *176*
Weighted least squares, recursive computation of, 131 ff.
Wersan, S. J., 65, *174*
Wide-sense conditional expectation, as optimal linear predictor, 166

Z

Zelen, J., 123, *174*
Zyskind, G., 90, 132, *176*